ABSOLUTE AND CHEMICAL ELECTRONEGATIVITY AND HARDNESS

ABSOLUTE AND CHEMICAL ELECTRONEGATIVITY AND HARDNESS

MIHAI V. PUTZ

Nova Science Publishers, Inc.
New York

Copyright © 2009 by Nova Science Publishers, Inc.

All rights reserved. No part of this book may be reproduced, stored in a retrieval system or transmitted in any form or by any means: electronic, electrostatic, magnetic, tape, mechanical photocopying, recording or otherwise without the written permission of the Publisher.

For permission to use material from this book please contact us:
Telephone 631-231-7269; Fax 631-231-8175
Web Site: http://www.novapublishers.com

NOTICE TO THE READER
The Publisher has taken reasonable care in the preparation of this book, but makes no expressed or implied warranty of any kind and assumes no responsibility for any errors or omissions. No liability is assumed for incidental or consequential damages in connection with or arising out of information contained in this book. The Publisher shall not be liable for any special, consequential, or exemplary damages resulting, in whole or in part, from the readers' use of, or reliance upon, this material.

Independent verification should be sought for any data, advice or recommendations contained in this book. In addition, no responsibility is assumed by the publisher for any injury and/or damage to persons or property arising from any methods, products, instructions, ideas or otherwise contained in this publication.

This publication is designed to provide accurate and authoritative information with regard to the subject matter covered herein. It is sold with the clear understanding that the Publisher is not engaged in rendering legal or any other professional services. If legal or any other expert assistance is required, the services of a competent person should be sought. FROM A DECLARATION OF PARTICIPANTS JOINTLY ADOPTED BY A COMMITTEE OF THE AMERICAN BAR ASSOCIATION AND A COMMITTEE OF PUBLISHERS.

LIBRARY OF CONGRESS CATALOGING-IN-PUBLICATION DATA

Library of Congress Cataloging-in-Publication Data

Putz, Mihai V.
 Absoute & chemical electronegativity & hardness / Mihai V. Putz.
 p. cm.
 ISBN 978-1-60456-937-7 (softcover)
 1. Electronegativity. 2. Acid-base chemistry. I. Title. II. Title: Absolute and chemical electronegativity and hardness.
 QD461.P98 2008
 541'.224--dc22
                    2008025688

Published by Nova Science Publishers, Inc. ✦ New York

CONTENTS

Preface		vii
Chapter 1	Introduction	1
Chapter 2	Basic Chemical Reactivity Principles	9
Chapter 3	Chemical Action Principle	19
Chapter 4	Systematic Electronegativity and Hardness	31
Chapter 5	Atomic Electronegativity and Hardness	45
Chapter 6	Molecular Electronegativity and Hardness	61
Chapter 7	Conclusion	75
Acknowledgments		79
References		81
Index		89

PREFACE

Systematic formulations of absolute and chemical electronegativity and hardness are analyzed among the local and non-local electronic density contributions in the frame of density functional theory. It is analytically proved that in all proposed cases can be founded the proper conditions within the absolute and chemical formulations to equalize. There appears that a new variational concept and term named as chemical action plays the unifying role among the quantum fluctuations of electronegativity and hardness at whatever level of atomic and molecular structural information. The power of these proofs consists in bypassing the knowledge of the total energy density functional. This way there was emerged out the new concepts of HOMO and LUMO chemical actions that neglecting the correlation-exchange terms account as the potential chemical works of the valence shells when exchanging electrons with the environment. As an application the associated atomic electronegativity, hardness and chemical action scales are computed and discussed for each unified quantum picture with the help of Slater orbitals. The so called bosonic electronegativity and hardness characterizing the fermionic-bosonic mixtures on valence states emerge out and their associate atomic scales are computed. It follows that they display periodic albeit inverse trends than those expected from pure fermionic behavior. This approach may be found most useful when explaining the Bose-Einstein condensates and superconductivity of atoms through electronegativity and hardness concepts. Extension to molecular systems is prospected by employing the recursive rules for electronegativity and hardness abstracted from electronegativity equalization principle combined with electronegativity-hardness invariant. In this molecular framework the unified forms of electronegativity and hardness are used to complete the proposed bonding scenario based on equality and inequality electronegativity and hardness reactivity principles for a specific series of Lewis bases. New index for checking the maximum hardness condition is formulated and applied as well. This way, the complete set of global electronegativity-hardness indicators of reactivity of atoms and molecules for various physico-chemical conditions is formulated in an elegant analytical manner within the conceptual density functional theory.

Keywords: electronegativity, chemical hardness, ionization potential, electronic affinity, density functional softness theory, chemical action functional, chemical action principle, electronegativity equalization principle, chemical potential inequality principle, hard and soft acids and bases principle, maximum hardness principle.

Chapter 1

1. INTRODUCTION

In an epistemological order the atomic and molecular constitutions are compressed by chemical reactivity; it is eventually followed by biological activity and social behavior.

However, the chemical reactivity stands as the bridge between the atomic and molecular levels of matter in their manifested state. In the continuous search for the quantitative description of the qualitative predictions of reactivity modern conceptual and computational chemistry had been raised [1-12]. Nevertheless, along the physico-mathematical models of atoms, molecules and clusters [10], relative simple indices yet with powerful chemical insight have been formulated in order to quantify the various levels of atomic organization of matter. In fact, among many formulations, the chemical history has retained those that are able to reflect the iterative nature of matter and bonds. This constraint roots from the former Dalton atomic theory prescribing that an atom has to preserve its most identity when combines with other forming molecular samples. In this context the electronegativity concept appears as one of the oldest vehicles in chemical prediction and interpretation of reactivity and bonding [3]. It emerges out from the Berzelius attempt to classify the substances as electropositive and electronegativity, under the strong influence of electricity concepts of his time [13]. From this point forward the concept of valence of an atom has been proved to give the proper description of the atoms-in-molecules combination emphasizing that the chemistry deals primarily with the outermost shells and orbitals of atoms and molecules [1, 14].

At this point worth to note that there was always opponents of assuming the virtual reality as the cause of the unfolded one - see [15] and the references herein. Enough recalling the famous enmities around the unification of space with time and of the fact that the time-space itself has a proper curvature as the Einstein theories prescribes [16]. In quantum physics the wave-function and then the orbital concept were blamed to lack the observable reality despite its proved probabilistic nature and confirmed spectral predictions. Remarkably, these days the wave-function concept was extended to the universe itself and the more deeply effects based on the so called hidden variables are employed in the current quantum information theories [17], leading with the teleportation and other virtual quantum or sub-space effects [18].

In quantum chemistry the electronegativity concept was highly disputed since the Pauling cornerstone definition of it as "a measure of the power of an atom in a molecule to attract electrons to itself" [14, 19]. From beginning, one could observe that this definition combines the atomic and molecular levels of a quality assumed to be the driving force for bonding, the electronegativity (χ or EN). However, to clarify the very meaning of this concept let's analyze it through the resonance structures:

$$A^{\delta-}B^{\delta+} \leftrightarrow AB \leftrightarrow A^{\delta+}B^{\delta-}, \tag{1}$$

giving the two limiting structures of the diatomic complex AB. Indeed, as the resonance structures are involved the power to attract electrons to itself seems somewhat an ambiguous sentence for atomic existence in a molecule.

In other words, there appears the fundamental question: which of the two bonded atoms of the reactions (1) is more electronegative? Aiming to solve such dilemma many survey reflections on the problem of electronegativity and related concepts have been put forth [20-26]. Regarding electronegativity Pauling's definition, by considering the atoms involved in one of the limiting structures of (1), $A^{\delta+}B^{\delta-}$ say, one could found two opposite phenomenological directions for the above question:

B is more electronegative since carriers more negative charge than its partner, or

A is more electronegative since its propensity to further attach extra-electronic charge to the molecule.

From the two opposite cases, both apparently true, the electronegativity paradox follows.

Still, this apparent paradox is, however, simple solved by the answer: none of them, while the most intuitive case requires that both electronegativities of atoms A and B in molecule AB equalize:

$$\chi_A = \chi_B \tag{2}$$

This way, without judging to literal the incipient assertion of Pauling, we have to recognize that his genius once again anticipated one of the most influential principles of chemistry: the electronegativity equalization principle (EEP) [27-50].

In favor of this interpretation and of principle of electronegativity we can invoke now also the simple and illumination perspective of Mulliken respecting the electronegativity of atoms-in-molecules [51]. He considered the reaction that corresponds to (1), replacing the limiting structures by the equivalent ionic components:

$$A^- + B^+ \leftrightarrow A + B \leftrightarrow A^+ + B^- \tag{3}$$

In this case, the relative tendencies of the two species to attract electrons can be quantified by the energies required to undergo the two side reactions of (3). This way, the energy required for that the reaction

$$A + B \rightarrow A^+ + B^- \tag{4}$$

to flow is given by the difference between the energy required to remove an electron from A, its ionization potential IP, and the energy consumed to attach the electron to the outer shell of B, its electron affinity EA:

$$\Delta E^{+-} = IP_A - EA_B \tag{5}$$

Similarly, for that the other equivalent reaction of (3) to be realized,

$$A + B \rightarrow A^- + B^+, \tag{6}$$

the consumed energy is

$$\Delta E^{-+} = IP_B - EA_A \qquad (7)$$

Now, as no preference between reaction (5) and (6) occurs in (3) their equivalency means the equalization of the involved energies, (5) with (7), respectively, which rearranges as:

$$IP_A + EA_A = IP_B + EA_B \qquad (8)$$

Equality (8) states in fact that a characteristic sum of each separate atom becomes equal with the correspondent one from other species when combine to form a bonded complex.

However, considering the semi-sum of the *IP* and *EA* quantities as qualitative definition of electronegativity for any species X (atom, molecule, or radical in its state of interaction) [51],

$$\chi_X = \frac{IP_X + EA_X}{2}, \qquad (9)$$

Mulliken inferred as well the electronegativity equalization principle (2), through (8) with (9).

There follows that the χ concept has to be seen as the *a priori* atomic property, in an *absolute* sense, that becomes *chemical* when atoms combine to form molecules or polyatomic systems. More, formally, the atomic and molecular χ has to have the same analytical expression, as fulfilled by (9), in agreement with the former Dalton intuition according with atoms chemically combines such that preserving most of its identity within a molecular system. Such demand allows for the recursive expansion (and explanation) of the world of substances and materials paralleling the complexity of the growing electron-nuclear collections of atoms. Therefore, concerning electronegativity, it has to account and can be seen as the average propensity of binding between two partners at whatever level of electronic-nuclear organization of matter.

From now, there is clear that having at hand concepts like electronegativity is of the most interest and helps when searching for predicting and explaining reactivity as a whole, since posing the virtually active properties of isolated atoms or groups of atoms

Before going to develop more accurately these kinds of global reactivity indices, let's make a note also for the units or dimension of electronegativity quantity. From definition (9) it seems to be of energetically nature although from Pauling definition it has to measure the "attracting power" of an atom. The conciliations between these aspects were given by assuming electronegativity as a potential, more precisely as the negative of the chemical potential μ of an electronic system [5, 52]:

$$\chi = -\mu = \left(\frac{\partial E}{\partial N}\right)_{V(x)}, \qquad (10)$$

being x the general spin-coordinate variable.

The definition (10) is without doubt a dynamical one: it regards the variation of the total energy E of an electronic system when the unit of electronic charge is transferred to the environment under the constraint that the potential under which the total N-electrons evolve, $V(x)$ - not restricted to the Coulomb one, remains constant. In other words, definition (10) provides an index for the spontaneous reactivity. More, it assumes electronegativity as a potential, from both physically and chemically perspectives.

There is however immediately that under assumption of the continuous and derivable energy function respecting total number of electrons, $E = E[N]$, the electronegativity of (10) approximates the former Mulliken formula (9) as the finite difference (FD) approach:

$$\chi = -\left(\frac{\partial E_N}{\partial N}\right)_{V(x)} \cong -\frac{E_{N_0+1} - E_{N_0-1}}{2} = \frac{(E_{N_0-1} - E_{N_0}) + (E_{N_0} - E_{N_0+1})}{2} = \frac{IP + EA}{2} \equiv \chi_{FD}, \qquad (11)$$

being N_0 the referential number of electrons in neutral state.

Of course, also here rises a problem, namely that of accepting a continuous distributions of energy through electric charges, to assure a context in which the derivation (11) is valid, a hypothesis hardly imagined by the classical electromagnetism. However, this also much disputed problem [53-58], finds solution within the quantum theory frame in which particles behaves also as waves, continuously by their very nature, being therefore no restricted to integer numbers when contributing to the total energy of a quantum system, as atoms and molecules are. This way, partial charges of atoms-in-molecules appears to be as well a reasonable effect of the quantum nature of the chemical bonding [52-62].

However, once clarified the electronegativity problem, through definition (11) and by its potential nature, there is clear that worth performing also the second derivative of the total energy respecting the total number of electrons, in the same conditions as equation (10) was considered:

$$\eta = -\frac{1}{2}\left(\frac{\partial \chi}{\partial N}\right)_{V(x)} = \frac{1}{2}\left(\frac{\partial^2 E}{\partial N^2}\right)_{V(x)}, \qquad (12)$$

introducing the so called chemical hardness [4, 24, 63-70].

At this point, another question deserves attention: why needs also the hardness along electronegativity to be conceptually considered? The answer relies on the physical nature of the electronegativity as potential. Since a complete physical picture of a phenomenon involves potential and forces there is straight that having a quantity with a potential nature in hand the other representing the corresponding force can be inferred by taking the minus of its gradient. The operation (12), grounded on the electronegativity concept (10) as the chemical potential of the concerned electronic system, fits well with this vision. Thus, the chemical hardness (12), in fact the quantity 2η, can be adequately called the associate *chemical force*. This interpretation as a chemical force perhaps motivates why in recent studies some groups prefer skipping the factor ½ form the basic definition of hardness (12) [71-74], albeit equally strong arguments are given bellow in the favor of maintaining it.

More, under the quantum assumption that the electronic charges behave also like waves as adopted, in finite approximation of the first derivative of the total energy E_N in (11), for chemical electronegativity, the respective expression for the hardness (12) can be as well laid down,

$$\eta = \frac{1}{2}\left(\frac{\partial^2 E_N}{\partial N^2}\right)_{V(x)} \cong \frac{E_{N_0+1} - 2E_{N_0} + E_{N_0-1}}{2} = \frac{IP - EA}{2} \equiv \eta_{FD}, \qquad (13)$$

in terms its own ionization potential and electron affinity.

This way, hardness, posses the same characteristics as its originator, the electronegativity, having the role of accompanying the reactivity global description from the force perspective so completing the driving reactive influence prescribed by the potential nature of electronegativity.

The last issue here has to be dedicated to necessity of the factors (1/2) included in the definition of electronegativity and hardness in expressions (9) and (12), respectively. This can be motivated in two ways.

One way is to consider that both electronegativity and hardness are regulated by the same occupation number q:

$$\chi = q(IP + EA),$$
$$\eta = q(IP - EA), \qquad (14)$$

due to the concerting effects that both indices assume, as approaching and establishing bonding, respectively. Now, in order that indeed χ and η to be affordable as parametrical minimal bond description one notes that the occupation number has to achieve the value $q=0.5$ as each atom within a bond contribute with one electron to its covalence. The non-integer value of q is in complete accordance with the quantum nature of the electron in bonds.

The second way is based on the equivalent gauge reactions, in which an acid-base complex is formed:

$$A^+ + {}^{\bullet}_{\bullet}B^- \leftrightarrow A^{\bullet}_{\bullet}B \leftrightarrow {}^{\bullet}A^- + {}^{\bullet}B^+ \qquad (15)$$

Eventually, the left reaction of (15) implies that partial charge transfer through Lewis acids and bases occur, while the right reaction of (15) states for the complete charge transfer redox process. In these conditions, the exchange of electrons between the radicalic extremes of (15) arises through the ionic/covalent - acid/base involved structures by means of general charge path:

$$\int_{N-1}^{N+1} dq = \int_{N-1}^{N} dq + \int_{N}^{N+1} dq = 1 + 1 = 2 \qquad (16)$$

The path (16) records at once the acidic (electron-accepting: $N \le q \le N+1$) and basic (electron-donating $N-1 \le q \le N$) behavior of the species, making it an inherent part of the description of their chemical reactivity [75]. Therefore, the electronegativity and hardness indices (14) have to be averaged by the charge path (16) leaving with the working semi-sum and semi-difference of the electron-releasing and electron-attaching energies, IP and EA, respectively [76-78].

We have therefore arrived at the every fruitful conception that chemical reactivity can be predicted and quantitatively characterized by global indices, end especially by electronegativity and hardness as ones of the most influential indices for the charge transfer and bonding. Additionally, we have agreed that only a quantum frame is the suitable picture in which the chemical reactions can be treated through chemical potential models. Finally, the definition of a suitable reactivity index, as electronegativity and hardness, has to preserve its formal structure at whatever level of atomic organization of matter, being *absolute* for the isolated atoms, groups and substances, and becoming *chemical* when they are analyzed in the state of interaction. More, the working reactivity indices have to pose the recursive character that is to can be logically (or intuitively) combined paralleling the complexity of bonding.

Certainly, all these ideas may seem strange for those who do not believe that the manifested reality has roots in virtual causes [15]; they are for sure limited to fail in searching of the very principles of nature only because a principle can not be directly measured but only proved by the effects prescribed. In this context, the orbital concepts, the aufbau principle of atomic order in the Periodic System [79-83], the molecular orbital theory [84-86], the density functional theory [87-95], the valence concept [1, 14, 19], or the chemical reactivity indices [96-120] and their principles [55, 57] are the specific tools for the conceptual and theoretical chemists that undergo the sacred mission of exploring the folded information of nature in order to better understand its flowering in acts and composites.

To be more plastic, to explore the chemical reactivity within quantum context is the same as Plato sought for the ideas of the manifested things and their transformations. In one day he was asked in agora: "maestro, how it is possible that I see the horse but not the idea of horse!" And the Plato replied: "this because you have not the eye with which to see that idea!"

The present chapter is primarily devoted to present the chemical reactivity as a whole, through global indices and their principles having electronegativity and hardness at the foreground. Electronegativity and electronegativity equalization principle are seen only as the first act in bonding and molecular formation. They are accompanied by the second effect, a sharper one, through the chemical hardness influence with a tuning role in stabilizing the molecular sample and bond through associated principles as well. Questing for different quantum formulation of electronegativity and hardness, and looking for their unification within a single analytical scheme for unfolding chemical reactivity stands as the main purpose of the actual venture and should become, when completed and validated, the holly grail of chemistry.

Chapter 2

2. BASIC CHEMICAL REACTIVITY PRINCIPLES

Although often arises the inquiry about χ and η physical reality, it can be assessed, for instance, by appealing to the already classical quantum picture of chemical bonding. There is well known that molecular bond is mainly characterized by bonding and anti-bonding orbitals, placed below and above of the non-bonding ones, respectively. This picture is well consecrated and its ability to explain many spectroscopic facts proved [121-129]. Performing now the phenomenological analogy, nevertheless a pertinent one, between the IP and bonding and between the EA and anti-bonding orbitals, appears that χ and η finely characterize the bonding nature as a whole, in an average sense of bonding and anti-bonding orbitals.

To be more specific, in the context of molecular orbital theory, according with Koopmans' theorem [130], the IP and EA can be written as the frontier orbital energies:

$$IP \cong -\varepsilon_{HOMO},$$
$$EA \cong -\varepsilon_{LUMO}, \qquad (17)$$

in terms of ε_{HOMO} and ε_{LUMO} as the eigenvalues of the highest and lowest occupied molecular orbitals, HOMO and LUMO, respectively. Replacing (17) in (11) and (13) χ and η acquire the computational expressions:

$$-\chi = \frac{\varepsilon_{LUMO} + \varepsilon_{HOMO}}{2},$$

$$\eta = \frac{\varepsilon_{LUMO} - \varepsilon_{HOMO}}{2}, \tag{18}$$

providing the route of particular evaluation without the recourse to calculation of the total energies of the electronic system, a laborious and not always very accurate task.

However, as far as the frontier orbitals are that ones mostly involved in chemical reactivity and bonding there follows that the role of χ and η through (18) in establishing the nature of the chemical bond is undoubtedly significant. They provide the most comprehensive principle of chemical reactivity as will be in next revealed.

Being about modelling reactivity, an elegant way of describing it calls the perturbation of the ground or valence state energy of an isolated system when engaging to an interaction [131, 132]. This is to write the energy as its expansion respecting to the changed charge ΔN into a specific reaction. At this point, even formerly such series expansion was considered up to the fourth order in ΔN [133-137], we consider that a simple and in principle complete scheme of global reactivity can be achieved on the electronegativity and hardness basis only, due to their character as the chemical potential and force, driving the changing in the total energy of the system, see relations (10) and (12), respectively.

In this context, the interaction energy $E_{\Delta N}$ of an electronic system that changes the charge ΔN with environment assumes the paradigmatic parabolic analytical form:

$$E_{\Delta N} = E_{0/v} + \mu_1 \Delta N + \eta_1 (\Delta N)^2, \tag{19}$$

standing for the total ground (subscript "0") or the valence (subscript "v") perturbed energy $E_{0,v}$ in the course of reaction through the chemical potential μ_1 and force η_1. Worth again noting, here from (19), the virtual nature of electronegativity and hardness as they are proper to a certain system in the absolute sense but becoming manifested chemical ones since the reaction flows, i.e. when $\Delta N \neq 0$.

Let's assume now that energy (19) is associated with the minimum perturbation to produce the chemical reaction or molecular (trans-) formation. Then, if further perturbation is considered respecting the change of charge, $\delta \Delta N$

with the small quantity $\delta \in [0,1]$, another reaction path is unfolding through the expansion:

$$\begin{aligned} E_{\delta\Delta N} &= E_{0/v} + \mu_1 \delta\Delta N + \eta_1 (\delta\Delta N)^2 \\ &= E_{0/v} + \mu_2 \Delta N + \eta_2 (\Delta N)^2, \end{aligned} \quad (20)$$

being $\mu_2 = \mu_1 \delta$ and $\eta_2 = \eta_1 \delta^2$ the new driving chemical potential and force, respectively.

Searching for the conditions for that the optimum reactivity path is naturally selected, we can immediately observe that as the variational principle of reactivity demands the restoring path from interaction energy (20) back to (19) one,

$$E_{\delta\Delta N} \to E_{\Delta N} \to \text{minimum}, \quad (21)$$

as the difference in their slopes and curvatures achieve maximum values. In these conditions, the variational principle for energetic paths is transferred to the variational principles at the level of chemical potential (or electronegativity) and chemical force (or hardness), accordingly written:

$$\Delta\mu = \Delta\mu_{\delta\Delta N \to \Delta N} = \mu_1 - \mu_2 = \mu_1(1-\delta) \geq 0, \quad (22)$$

$$\Delta\eta = \Delta\eta_{\delta\Delta N \to \Delta N} = \eta_1 - \eta_2 = \eta_1(1-\delta^2) \geq 0 \quad (23)$$

This way, the chemical reactivity principles emerge from the exploitation of the electronegativity and hardness principles, (22) and (23), respectively. These are known as chemical inequality and equality (or neutralization) principles when (22) is employed as inequality or equality, respectively, while the hard-and-soft acids and bases and maximum hardness principles are obtained from equality and inequality considerations of (23), respectively. They are also summarized in the table 1.

One has to bear in mind that both principles (22) and (23), with the special principles of table 1, appear when bonding or a chemical reaction is concerned, and a comprehensive theoretical analysis has to include all these causes in interpreting the reactivity effects.

Table 1. Synopsis of the basic principles of reactivity, at the chemical potential (or electronegativity) and chemical force (or hardness) levels, as abstracted from analytical principles (22) and (23), respectively

Bonding Index	General Principle	Special Principle	Principle of Bonding
$\mu = -\chi$	$\Delta\mu \geq 0$	$\Delta\mu = 0$	Chemical potential (or electronegativity) equality (EE): "the chemical potential of all constituent atoms in a bond or molecule have the same value" [5]
		$\Delta\mu > 0$	Chemical potential (or electronegativity) inequality (EI): "the constancy of the chemical potential is perturbed by the electrons of bonds bringing about a finite difference in regional chemical potential even after - chemical equilibrium is attained globally" [137-140]
η	$\Delta\eta \geq 0$	$\Delta\eta = 0$	Hard-and-soft acids and bases (HSAB): "hard likes hard and soft likes soft" [141-145]
		$\Delta\eta > 0$	Maximum hardness (MH): "molecules arranges themselves as to be as hard as possible" [116, 117, 119, 120]

Therefore, the global scenario of reactivity, based on electronegativity and hardness principles of table 1, implies that there are four stages of bonding:

i. approaching stage is dominated by the difference in electronegativity between reactants and is consumed when the electronegativity equalization principle is fulfilled among all constituents of the products; this stage is associated with the charge flow from the more electronegativity regions to the lower electronegativity regions in a molecular formation thus covering the *covalent* binding step;

ii. even after the chemical equilibrium is attained globally the electrons involved in bonds acts as foreign objects between pairs of regions, at whatever level of molecular partitioning procedure, inducing the appearance of finite difference in adjacent electronegativity of neighbor regions in molecule; it is due to the quantum fluctuations associate with the quantum nature of the bonding electrons and it corresponds to the degree of *ionicity* occurred in bonds;

iii. the induced ionicity character of bonds is partially compensated by the chemical forces through the hardness equalization between the pair regions in molecule; the HSAB principles is therefore involved, as a second order effect in charge transfer – see expansion (19) for instance, being driven by the ionic interaction through bonds;
iv. still, the quantum fluctuations provides a further amount of finite difference, this time in attained global hardness, that is transposed in relaxation effects among the nuclear and electronic distributions so that the remaining unsaturated chemical forces to be dispersed by stabilization of the molecular structure.

This way, there was proved that electronegativity and hardness provides the minimum set of reactivity indices able to cover the complete process of binding, as a whole.

In order to make a more intuitive idea how above principle act at the energetic level the figure 1 depicts the equality and inequality variants of them, in (a) and (b), respectively.

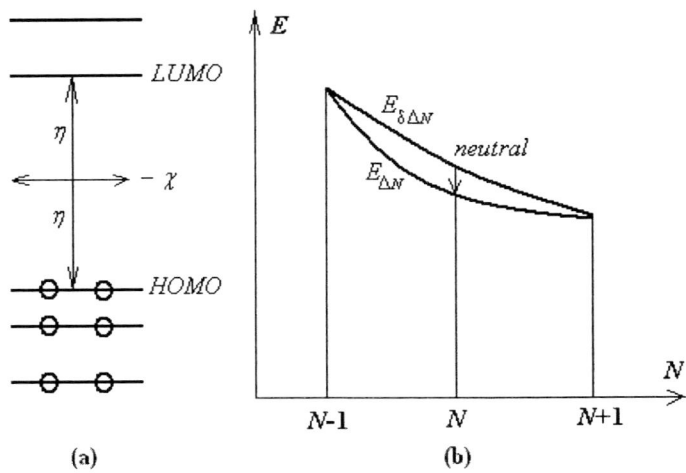

Figure 1. (a) Orbital energy diagram for a molecule [143], showing the electronegativity and hardness according with (18), on which basis the equalization of electronegativity and hardness principles, EE and HSAB of table 1, follows. (b) Plot of the electronic energy vs. electrons for a molecule [55], on which base the electronegativity inequality and maximum hardness principles of chemical reactivity, EI and MH of table 1, follow.

From the figure 1(a) there is immediately that during a reaction or bonding the difference in electronegativities of the partners is encountered firstly, as the main effect of reactivity will be the adjustment of the middle of the HOMO-LUMO gap so that easier frontier transfer of the electrons to be realized. This explains the EE principle of table 1 on an energetic relative scale.

More, one can see that hard molecules are characterized by large HOMO-LUMO gap while soft molecule by a small one; therefore, as a second effect of reactivity, the hard-hard and soft-soft interactions are favorite respecting the hard-soft and soft-hard ones since the exchange of the charge induced by the ionic character of bonding is much easier performed through the first two cases due to the close lying of the energies of the frontier orbitals. This explains also the HSAB principle on a relative energetic scale.

Instead, from figure 1(b) is getting out that as the energetic curve of interaction, with the paradigmatic parabolic form (19), approaches the optimum reaction path, to achieve its minimum for the neutral collection of nuclei, the slope ($\Delta\mu$) and the curvature ($\Delta\eta$) of the energetic limits (21) reach their maximum, that is the graphical formulation of the chemical potential and hardness inequality principles, recovering the IE and MH reactivity principles of table 1.

Nevertheless, for practical purposes a closer quantitative rationalization of the chemical reactivity principles in terms of electronegativity and hardness is desirable. As such, regarding the electronegativity equalization principle many studies are performed with considerable successful results by making at work the interaction energy of type (19) through expressions (2) and (10) for a large collection of compounds and reactions [36-41]. However, the situation is not outstanding when hardness is called through its associate MH and HSAB principles to quantify the molecular formations and description of chemical reactions [146-156]. The existing debates are founded on the fact that still is questing for an adequate quantity with which hardness to correlate when a propensity for bonding is studied. There are studies that perform a parallel analysis of hardness variation with the exchanged charge ΔN [99, 142], while others are done respecting the reaction energies ΔG [74], with relative meaningful results. More tedious, statistical correlations between the enthalpies of the Lewis acid-base interactions were also performed to include the electrostatic, covalent and even the so called transmittance-receptance terms of the transferred electron charge during reactions [157].

There is quite surprising that after our best knowledge no systematic studies are reported for linking the hardness with its conceptual source, the electronegativity when applying the HSAB principle. Such link is therefore here advanced based on the very definitions of what soft and hard acid and base are. Still, the right connection can be achieved recalling that electronegativity has the potential nature at the chemical level, so being proportional with the inverse of the radius of atoms or length of bonds, $\chi \propto 1/r$. Worth noting that such dependence was the main picture in which one of the most recent atomic electronegativity scale was given [83]. It can be equally derived from a simple model of charging energy of a conducting sphere of radius r [81, 97, 142]. Yet, electronegativity can be seen as being proportional also with the inverse of the polarizability, $\chi \propto 1/r$, since polarizability α is on its turn proportional with the volume encompassed by the electronic system under discussion [82]. With these remarks in hand let's list the main definitions of soft and hard acids and bases, connecting their hardness degree with electronegativity:

- a soft base, e.g. R^- or H^-, is very polarizable and thus with low electronegativity;
- a hard base, e.g. F^- or OH^-, is not much polarizable and thus with high electronegativity;
- a soft acid, e.g. RO^+ or HO^+, has usually low positive charge and large size, so posing lower electronegativity;
- a hard acid, e.g. H^+ or XH (hydrogen bonding molecules), has normally high positive charge and small size, so posing high electronegativity.

More, one can straightforwardly infer from figure 1 (a) that the relative position of electronegativities between two reactants can give the acid or base nature of the species since the more basicity the more $-\chi$ is pushed towards positive range. This observation seems crucial to us and can explain why the consecrated classification of some common compounds to be acids or bases [4, 68] has not an absolute value and founds some computational disagreements [74, 151], while the relative electronegativities involved in concerned reactions have to as well be taken as the appropriate measure.

Collecting all these ideas in a representative quantum concept we can draw the figure 2 for appropriate indication of the acid/base and hard/soft trends of the chemical species within the electronegativity-hardness chemical space (χ, η). The figure 2 (a) depicts the phenomenological correlation between electronegativity and hardness for a given chemical species leaving with their natural classification as acids and bases on the electronegativity scale and hard and soft on the hardness scale as their positions are more departed from the (0, 0) origin point within the chemical space (χ, η). A similar classification can be done in figure 2 (b) for a series of acids and bases, separately, with the result in categorizing them as hard-and-soft character with low-and-high polarity as their positions are more departed from the (0, 0) origin point within the chemical space (χ, η). This way, there is provided a new valuable working scheme with the help of which the relative acidic or basic nature as well as the hard and soft strength of the molecules in their state of interaction is analyzed.

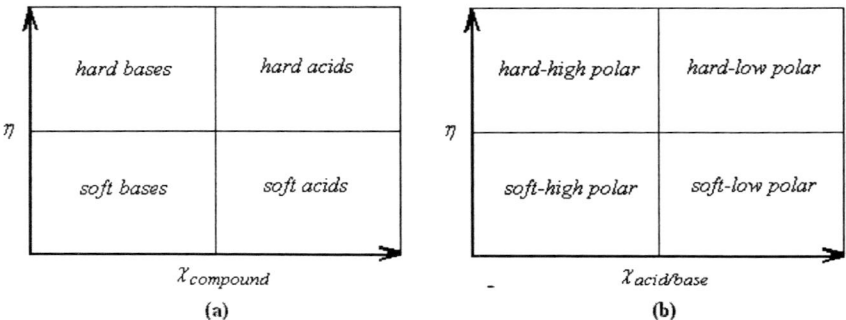

Figure 2. (a) The diagram of the compound repartition as hard-and-soft acids and bases within the electronegativity-hardness space (χ, η). (b) The diagram of the hard-and-soft nature of the high-and-low polar acids and bases within the electronegativity-hardness space (χ, η).

Having from now a rationale for the HSAB principle in term of electronegativity, through the (χ, η) chemical space, still remains to suitable quantify the MH principle. Also here, despite much theoretical concern for its proof [117, 119] and of the important computational confirmations [118, 121], there seems that the proper index to measure the degree in which the maximum hardness is achieved or not in certain circumstances is still missing. Going to fill this lack let's recall from table 1 and of the related bonding stages that the MH principle states at the end of reactivity effects, employing the rest of the uncompensated hardness in bonds to stabilize the molecule. Therefore the MH principle involves a kind of self-difference of hardness. However, it has to characterize at the same time the resulting global hardness obtained through the hardness equalization (or HSAB step) from the preceding stage in bonding. From here the delicate point to have, in principle, a single hardness at the end of HSAB stage and to be as well involved in a sort of internal difference targeting molecular stabilization and measuring the resistance to further combinations. This situation can be elegantly solved if the inverse of hardness is introduced, the softness [5, 55,104, 105]:

$$S = \frac{1}{2\eta} \qquad (24)$$

Worth noting that the softness achieves the conceptual inverse meaning of hardness, namely the propensity of engaging into a reaction, leaving with the appropriate possibility of reformulation of the above hardness principles of bonding [55-58]. Due its characteristic that directly relates polarizability as well [109], global softness is often more conveyable to deal with when describing reactivity.

More, once having introduced softness as inverse of hardness, and measuring at the same time the degree of polarizability, one can also remark from figure 2(b) that the hard-high, hard-low, soft-high, and soft-low polar classification recovers the hard-soft, hard-hard, soft-soft, and soft-hard conventional taxonomy of acids and bases within the (χ, η) space, respectively.

However, here, as MH principle follows the HSAB principle the hardness and softness combination looks the best way to be considered. This way, in order to quantify the final stage of bonding the MH index is proposed:

$$Y = \frac{\eta - S}{\eta} = 1 - \frac{S}{\eta} \stackrel{(24)}{=} 1 - \frac{1}{2\eta^2}, \qquad (25)$$

that, obviously, closely tends to its maximum value, $Y = 1$, when the hardness is as maximum as also its softness (24) tends to zero.

The index (25) was called with Greek "Y" because is the closest available – not attributed yet - majuscule to the Geek "I" of η echoing for that reason the "maximum hardness". It is constructed as a difference on the global hardness value acquired via hardness equalization of the HSAB bonding stage.

To better emphasize on the maximum hardness index (25) in the context of hard-and-soft reactivity worth rewriting it in an explicit manner as:

$$Y = \frac{\eta}{\eta} - \frac{S}{\eta} \qquad (26)$$

Expression (26) is particularly relevant for MH principle because lights on how the molecular stability is related with the difference between the hard-hard (η / η) and soft-hard (S / η) ratios within the hard (shorthanded as "h") and soft (shorthanded as "s") acids and bases reactions,

$$h_1 - s_1 + s_2 - h_2 \leftrightarrow h_1 - h_2 + s_2 - s_1, \tag{27}$$

during which a soft base (s_1) bonded to a hard acid (h_1) removes a hard base (h_2) from a soft acid (s_2). As such, the difference (26) measures the degree with which the equilibrium in (27), say between $h_1 - h_2$ and $s_2 - h_2$, is broken to favor or not the stabilization of hard-hard and soft-soft products.

Remarkably, the MH index (25) allows the existence of two cases of reactivity. One is when $Y \in [0,1]$ which indicates that the departure of the equilibrium in (27) takes place to its right side as (25) closely tends to 1, leaving with the meaning that hard-hard bond is more favorable than the soft-hard state of adducts ($h - h \gg s - h$, or $\eta/\eta \gg S/\eta$). The other case arises when $Y < 0$ in (25) thus indicating that the stabilization process is not finished respecting HSAB principle: as far as $\eta/\eta < S/\eta$ implies that $h - h$ bond is less strong respecting $s - h$ ones, leaving the equilibrium in (27) more shifted to its left side.

Overall, throughout this section was clearly illustrated how the electronegativity and hardness concepts can provide simple yet insightful frame for treating chemical reactivity from their associate principles. They will be in next joined with another useful concept and principle with a quantum role in unification of their different analytical formulations within one of the most preeminent quantum theory of nature, the Density Functional Theory (DFT).

Chapter 3

3. CHEMICAL ACTION PRINCIPLE

With the birth of quantum chemistry as the new conceptual frame for physical chemistry, appears also the need of the in deep characterization of the chemical concepts with the physico-mathematical tools of quantum mechanics. With this desideratum the quantum theory of atoms and molecules were firstly formulated by Pauling, Mulliken, and Slater, employing the already classical wave function or orbital theory of matter [19, 68]. However, despite of the impressive success in explaining and predicting reactivity the wave function representation suffers for enlarging to much the dimension of the reactivity space to that of the Hilbert-Banach space.

Fortunately, even not less abstract, alternative representation of quantum constitution of matter was emerging from the incipient quantum statistical works of Thomas, Fermi and Sommerfeld until the lucid density functional theory of the many-electronic systems due to Hohenberg, Kohn, and Sham [87, 88].

The result was a new quantum vision of nature, the Density Functional Theory (*DFT*) [158-160], leading with a whole rewriting of the consecrated concepts in physical chemistry, from valence and bonding until the stereo-selectivity and reactivity ones [161-166].

This effort covers the last five decades in quantum chemistry research and is due to the impressive abstract yet chemical intuitions of Parr, Pearson, and People schools of physical and mathematical chemistry. They lead with the new paradigm of treating the structure of atoms and molecules and of their correlations with the manifested chemical properties. Not less important, the electronegativity and hardness concepts have found their combined qualitative-quantitative realization as well their associate principle of bonding, i.e. EE, EI, HSAB, and MH principles of table 1.

The key passage from the electronic wave function to electronic density was done though understanding that the quantification of concepts means quantification of their functions releasing with the functional framework. As a consequence, in modern physical chemistry the quantum numbers are density functionals, i.e. functions of functions, assuring therefore that an entire structure or process is counted when projected or sublimated in a quantitative analysis.

However, in what follows the main theorems and flavor of *DFT* are presented through presenting the Hohenberg-Kohn theorems, emphasizing on the limiting threshold that still persists in knowing the total energy of a system as an explicit density functional expression. Nevertheless, as often happened in development of quantum chemistry, this, in principle complete, quantum picture allows retaining of the significant part for reactivity, the potential of the electron interaction with a collection of nuclei, in a form of density functional quantity called chemical action [12, 167].

The associate derived principle, its use in chemical reactivity, and connections with electronegativity and hardness are as well consequential and contributes in unifying a strongly debated issue [3, 5, 134, 135], the orbital with the global nature of the global indices and bonding.

3.1. HOHENBERG-KOHN THEOREMS

In short, Density Functional Theory is based upon two fundamental principles known as Hohenberg-Kohn (*HK*) theorems [87].

(i)   The true state electronic density $\rho(x)$ determines everything about a chemical (many-electronic) system and integrates to the total number of electrons in the system:

$$\int \rho(x)dx = N \tag{28}$$

Relation (29) provides, however, an entire revolution of the old quantum chemistry thinking since replaces the normalization condition of the wavefunction. In the present acceptation the abstract one-electron-*N*-variables-wavefunction becomes *N*-electrons-single-variable-density with a clear more intuitive representation of the poly-electronic systems. It is accompanied by the total energy representation as the electronic density functional of which principle is stetted by the second theorem of *DFT*.

(ii) For any trial electron density $\overline{\rho}$ holds the variational principle of total energy in *DFT*, around the true state density ρ.

$$E[\overline{\rho}] \geq E[\rho] \Leftrightarrow \delta E[\rho] = 0, \qquad (29)$$

Here, another comment is required as well. For instance, the "total energy functional" nomination seems somehow confusing as no single number can explain the wide energy spectrum. In this respect one has to understand that the electronic density ρ is not unique for a system, varying from eigen-state to eigen-state. Still, this formulation allows the freedom of considering the ground as well the valence state alone in quantum considerations concerning reactivity since the conservation of charge in that state is assured though first *HK* theorem (28).

As such, the both *HK* theorems can be combined to provide the working energetic variational principle of the eigen-energy corresponding to a true electronic density ρ characterizing the distribution of *N*-electrons in that state. It has the form [5]:

$$\delta\{E[\rho] - \mu N[\rho]\} = 0, \qquad (30)$$

being μ the Lagrange parameter that caries the role of adjusting the energy surface so that to become minimum, as (29) requires, for a given state within (28) constrain.

There is therefore most striking that from the very quantum principle of energy raises the chemical potential (or the electronegativity) existence as the key parameter in minimizing it, posing from (30) the functional derivative definition:

$$\chi = -\mu = -\left(\frac{\delta E[\rho]}{\delta \rho}\right)_{\rho = \rho(V)} \qquad (31)$$

The link with the previous conceptual definition (10) is quite straight through the series of identities [5]:

$$\chi = \chi\left(\frac{\partial N}{\partial N}\right)_V \overset{(28)}{=} \int \chi\left(\frac{\partial \rho}{\partial N}\right)_V dx \overset{(31)}{=} -\int \left(\frac{\delta E}{\delta \rho}\right)_V \left(\frac{\partial \rho}{\partial N}\right)_V dx = -\left(\frac{\partial E}{\partial N}\right)_V \qquad (32)$$

There is therefore proved that electronegativity finds its proper place in the *DFT*, so casting the promise to deliver the most comprehensive analytical expressions, density functionals, and quantification of the chemical reactivity.

3.2. CHEMICAL ACTION PRINCIPLE THROUGH TOTAL ENERGY PRINCIPLE

From both theoretically and experimental point of view, the most important density functional stands the total energy of a system evolving under the external potential $V(x)$. It separates as [88]:

$$E[\rho] = F_{HK}[\rho] + C_A[\rho], \qquad (33)$$

where

$$F_{HK}[\rho] = T[\rho] + V_{ee}[\rho] \qquad (34)$$

represents the *HK* functional, written as the sum of the electron kinetic functional $T[\rho]$ and the electron repulsion functional $V_{ee}[\rho]$, respectively, whereas the exact term:

$$C_A[\rho] = \int \rho(x) V(x) dx \qquad (35)$$

represents the *chemical action* [167] for the reasons bellow revealed.

Unfortunately, despite its formal simplicity the energy functional is still an unknown exact analytical expression, although many approximations and models have been proposed for [6]. Worth noting that depending on the level for approximating *HK* functional (34) the resulted computational schemes leave different numbers for quantifying the same reactivity effects, creating in certain circumstances embarrassment in chemical interpretation [8]. However, especially when dealing with chemical reactivity one can restrict the analysis to the valence effects only, without being afraid of loosing interpretation and much of the quantitative of the involved phenomena.

Therefore, based on the universal nature of the unknown functional (34), the variational principle for total energy can be decomposed as well:

$$\delta E[\rho] = 0 \Leftrightarrow \delta F_{HK}[\rho] + \delta C_A[\rho] = 0 \qquad (36)$$

providing that both components obey, separately, the same variational constraint:

$$\delta F_{HK}[\rho] = 0, \qquad (37)$$
$$\delta C_A[\rho] = 0 \qquad (38)$$

Since the optimum principle at the HK level (37) furnishes the information regarding the electronic repulsion, exchange and correlations energies, in short of inner effects among the electrons on different states and shells of an atom or a molecule, being a subject of approximation and relative quantification, the chemical action level (38) allows full treatment of the effects occurred due of the external (or nuclear) influence upon the electronic system. However, it delivers best results when electronic density in (35) is restricted to the valence shells or orbitals.

This way, because the exact analytical form of the functional $F_{HK}[\rho]$ is not yet known the chemical action $C_A[\rho]$ should assumes a fundamental role in the structure and modification of the many-electronic valence states within DFT framework.

However, worth to note that it was Mel Levy the first one who, using the universal properties of the Hohenberg-Kohn functional, had arrived to a version of the above chemical action principle, available both for an arbitrarily large M-set of non-interacting as well as of interacting Hamiltonians, through the minimization of his G functional [168]:

$$G_{1,2,...,M}^{\alpha,\beta,...,\omega} = \int dx \left[\rho_\alpha(x) V_1(x) + \rho_\beta(x) V_2(x) + ... + \rho_\omega(x) V_M(x) \right] \qquad (39)$$

The minimum of functional (39) establishes in fact a realization of the chemical action principle (38) and optimizes the ordering pairs of the densities with the associate external potentials.

More, the chemical action concept and its principle fit with the recommendation of the HK theorems from which the external potential and the electronic density are uniquely correlated for a given eigen-state. In this property resides the immense potentiality for the practicing of this concept to derive and control the density functionals, i.e. to quantify the reactivity. An illustration of the chemical action conceptual reliability will be later presented regarding the electronegativity and hardness density functional formulations.

3.3. CHEMICAL ACTION THROUGH TOTAL ENERGY [N, V] REPRESENTATION

One of the most important issues in *DFT* regards the *N*- and *V(x)*-representability of the working electronic density that is used to construct the global density functionals, in particular the total energy [92]. While *N*-representability properly associates a trial density with the integral constrain (28) the *V(x)*-representability problem consists in finding the right density that uniquely correlates with the overall potential applied to the electronic system or in which the electronic system evolves [93]. The last condition is usually achieved through fulfilling the second *HK* theorem (29), usually under the (30) form.

As such, in order to see which unified condition has to satisfies both *N*- and *V(x)*-representability constraints worth expanding the change in total energy functional of a system onto the restricted base [*N, V(x)*], that restricted to the first order expansion looks like [158]:

$$dE = \mu dN + \int \rho(x) dV(x) dx \qquad (40)$$

From (40) one can immediately recognize that the controlling parameter and function of this expansion are the chemical potential and the electronic density:

$$\mu = \left(\frac{\partial E}{\partial N} \right)_{V(x)}, \qquad (41)$$

$$\rho(x) = \left(\frac{\delta E}{\delta V(x)} \right)_{N}, \qquad (42)$$

respectively, since interpreting the total energy from the thermodynamical potential perspective. However, to link the change of the total energy (40) with the energy variational principle recommended by the second *HK* theorem worth rearranging it as:

$$dE - \mu dN = \int \rho(x) dV(x) dx, \qquad (43)$$

firstly, and then taking the path integral over the reaction coordinate, followed by the functional differentiation around the eigen-states to assure the minimization procedure. There results the equivalent form of (43) that is:

$$\delta\left\{\int [dE - \mu dN]\right\} = \delta\left\{\int \left[\int \rho(x) dV(x) dx\right]\right\} \qquad (44)$$

Now, recalling that around eigen-states the chemical potential (or electronegativity) is essentially constant, due to the chemical potential or electronegativity equalization principle (EE) of table 1,

$$\mu = CONSTANT, \qquad (45)$$

and performing the integrations along the considered reaction path in (44) there leaves with the variational identity:

$$\delta\{E[\rho] - \mu N[\rho]\} = \delta C_A \qquad (46)$$

that appears to be the most general relationship between chemical potential (negative of electronegativity) and the total energy, through the chemical action [112].

At this point there is evident that since replacing the charge density functional by its definition (28) in (46) there results that the chemical action alone sustains both the *HK* theorems, by considering also of the energy minimization (30), through its associate optimum principle:

$$\delta C_A = \delta\left[\int \rho(x) dV(x) dx\right] = 0 \qquad (47)$$

with this re-derivation of the chemical action principle, the chemical action concept acquires first of its interpretation: as far as the left hand side of equation (46) establishes the *stationary principle* in *DFT*, the right hand side formally recovers the *variational principle* for the system's action. This way, C_A of (35) assumes an *action* meaning, here at the chemical level.

Further interpretations of the chemical action, together with its functional relation with electronegativity and hardness, are in next presented.

3.4. CHEMICAL ACTION CONCEPT AND ITS FUNCTIONALS

Aiming to understand the role of the chemical action in characterizing the atomic and molecular samples one can immediately remark that the functional form (35) can be seen as the average of the external potential over the electronic density of a certain state. In fact, chemical action reflects the observable potential of an electronic state:

$$C_A = \langle V(x) \rangle_{\rho(x)} = \int \rho(x) V(x) dx. \tag{48}$$

with the observable nature of the chemical potential we can bring it to the virial theorem in writing the total energy of atoms and molecules [169, 170]:

$$E^{at} = 0.5(C_A + \langle V_{ee} \rangle),$$
$$E^{mol} = 0.5(C_A + \langle V_{ee} \rangle + \langle V_{nn} \rangle), \tag{49}$$

where $\langle V_{ee} \rangle$ and $\langle V_{nn} \rangle$ are the electronic and nuclear interaction energies, respectively.

Now, keeping only electron-nuclear interaction energies together with the other external energies applied to the system, we are situated in the same level of approximation as the frame in which the chemical action principle (38) was deduced via neglecting of the universal *HK* functional terms (34). This can be justified because of their universal natures, since kinetic energy of a free electron and the repulsion energy between two electrons have to have the same values whatever the external potential is applied upon them; thus treating these energies as additive constants to the total energy of a system equally they can be subtracted; this way, the remaining terms account for the particular energetic behavior depending of the strength and type of the existing external potential, quantified by means of chemical action hereafter. However, at the molecular level, due to the screening of the electronic clouds the subtraction of the nuclear-nuclear repulsion can be as well admitted at a certain level of approximation. In these conditions, one can roughly see the chemical action as double quantity of the total energy of an electronic system when the electronic universal energetic terms as well as the nuclear ones due the screening effects are omitted:

$$C_A^{at,mol} \approx 2E^{*at,mol} \qquad (50)$$

More, in the same phenomenological context an orbital interpretation of the chemical action can be as well presented. It is based on rewriting of the integral definition, (35) or (48), as the sum of the orbital chemical actions (c_{Ai}) of all orbitals laying bellow that of the current averaged state density:

$$C_A = \int \rho_N(x)V(x)dx \cong \sum_{N_i}^{N} \int \rho_{N_i}(x_i)v(x_i)dx_i = \sum_{N_i}^{N} \langle v(x_i) \rangle = \sum_{N_i}^{N} c_{Ai} \qquad (51)$$

This way, chemical action can also be seen as the double sum of all orbital energetic contributions,

$$C_A \approx 2\sum_i \varepsilon_i^*, \qquad (52)$$

since applying the modified virial theorem at the orbital level when the electron kinetic, electron-electron and nuclear-nuclear energies are omitted:

$$\varepsilon_i^* \approx 0.5 \langle v(x_i) \rangle_{\rho_{N_i}(x_i)} \qquad (53)$$

with these approximations that reveal many faces of the chemical action concept, let's light in next also on the effects of the chemical action principle, (38) or (47). For that, performing the explicit functional derivation respecting the electronic density and external potential the chemical action principle rewrites as:

$$\delta C_A[\rho] = 0 \Leftrightarrow \int \rho(x)[-\delta V(x)]dx = \int \delta \rho(x)V(x)dx \qquad (54)$$

Aiming to give significance of the terms of (54), one can easily observe that the right hand side of (54) covers the average of the change in *local chemical work* (δC_W) induced by the *local chemical force*, $\vec{f}_V(x)$, due to its relation with the minus of the functional gradient of external potential:

$$\langle \delta C_W \rangle = \int \rho(x) \left[\vec{f}_V(x) \cdot \vec{\delta x} \right] dx = \int \rho(x)[-\delta V(x)]dx \qquad (55)$$

Thus, by combining (54) with (55) there is inferred that the chemical action achieves also the meaning of the averaged chemical work performed on or from the concerned system:

$$C_A = \langle C_W \rangle = \int \rho(x) V(x) dx \qquad (56)$$

From this last quality the chemical action concept relates directly also with the electronic affinity and ionization potential, when the system performs a chemical work to attach or to detach an electron form its current state, respectively:

$$C_A^{+1} = \int \rho_{N+1}(x) V(x) dx \equiv EA_{CA} \cong -\varepsilon_{LUMO-CA}, \qquad (57)$$

$$C_A^{-1} = \int \rho_{N-1}(x) V(x) dx \equiv IP_{CA} \cong -\varepsilon_{HOMO-CA}, \qquad (58)$$

within the context of vertical electronic transfers so that the Koopmans' theorem can be also applied to release the chemical action LUMO and HOMO related orbital energies.

With respective chemical action chemical affinity and ionization potential of (57) and (58), the associate chemical action electronegativity and hardness can be formulated by combining them in the working expressions (11) and (13), respectively:

$$\chi_{CA} = \frac{C_A^{-1} + C_A^{+1}}{2}, \qquad (59)$$

$$\eta_{CA} = \frac{C_A^{-1} - C_A^{+1}}{2} \qquad (60)$$

The importance of introducing pure chemical action electronegativity and hardness consists – firstly, in replacing the total energy framework with the exclusive external potential based one. Then, applying the virial approximation to the chemical action functionals (57) and (58), as in (51) with (52),

$$C_A^+ = \int \rho_{N+1}(x) V(x) dx \cong \sum_{N_i}^{N} \int \rho_{N_i+1}(x_i) v(x_i) dx_i \approx 2 \sum_{N_i}^{N} \varepsilon_i^{+*}, \qquad (61)$$

$$C_A^- = \int \rho_{N-1}(x)V(x)dx \cong \sum_{N_i}^{N} \int \rho_{N_i-1}(x_i)v(x_i)dx_i \approx 2\sum_{N_i}^{N} \varepsilon_i^{-*}, \qquad (62)$$

one gets the global-orbital electronegativity and hardness relationships, respectively as:

$$\chi_{CA} = \sum_i \chi_{CAi}, \qquad (63)$$

$$\eta_{CA} = \sum_i \eta_{CAi}, \qquad (64)$$

while the orbital electronegativity and hardness assumes the orbital energetic definitions:

$$\chi_{CAi} \approx \varepsilon_i^{-*} + \varepsilon_i^{+*}, \qquad (65)$$

$$\eta_{CAi} \approx \varepsilon_i^{-*} - \varepsilon_i^{+*} \qquad (66)$$

The actual results have the merit to unify the orbital and global pictures of electronegativity and hardness, a matter being long time under discussions [3, 5]. There was indicated that orbital quantities sum up to give the global ones when neglecting all the pure electronic and nuclear energetic interactions. Otherwise they have to become all equal, and equal with the global representative quantity as prescribed by the equalization electronegativity (EE) and hardness (HSAB) principles of bonding, see table 1. However, chemical action concept manifests as a flexible theoretical tool bridging total energy, electronic density, chemical work and reactivity indices within a single unitary framework in which the external potential gives the dominant character of analysis. This approach becomes most useful when only valence shells are invoked to modeling and interpreting chemical reactions. However, in what follows the chemical action will be always present, in a direct or indirect manner, to characterize and further contribute to unify various density functionals formulations of electronegativity and hardness, assuring so far the way of unifying the orbital with the global levels of reactivity.

Chapter 4

4. SYSTEMATIC ELECTRONEGATIVITY AND HARDNESS

In the context of *DFT* different electronegativity and hardness functionals are derived, within different absolute or chemical pictures, employing the local or non-local or both combined properties of softness hierarchy. At each level of approximation or comprehension the chemical action or its related terms are involved with the unification role over the implicit orbitals, as previous discussed.

4.1. ABSOLUTE AND CHEMICAL ELECTRONEGATIVITY AND HARDNESS

For an *N*-electronic system, being under the external potential influence $V(x)$, its functional energy $E = E[N, V(x)]$, admits the total differential equation with the form (40).

Nevertheless, by using the Cauchy property of the total differentiable quantities the Maxwell type relation can be associated with the equation (40):

$$-\left(\frac{\delta \chi}{\delta V(x)}\right)_N = \left(\frac{\partial \rho(x)}{\partial N}\right)_V, \qquad (67)$$

where the chemical potential was replaced by its minus *absolute* electronegativity (10).

Now, being Δ the change in electrons, restricted to the valence shell, equation 40 can be rearranged to be integrated successively as:

$$E(N_v \pm \Delta) - E(N_v) = - \int_{N_v}^{N_v \pm \Delta} \chi dN + \int \rho(x)\{\int \delta[V(x)]\}_{N_v} dx$$

$$= - \int_{N_v}^{N_v \pm \Delta} \chi dN + \int \rho_{N_v}(x) V_{N_v}(x) dx, \qquad (68)$$

around the number of concerned valence electrons, N_v.

This way, in equation (68) appears the influence of the chemical action, (35) or (48).

Next, assuming the variation of electronic charge in (68) to be unity ($\Delta = 1$) the electronic affinity and ionization potential can be recovered to be respectively:

$$-EA = - \int_{N_v}^{N_v+1} \chi dN + C_A,$$

$$IP = - \int_{N_v}^{N_v-1} \chi dN + C_A \qquad (69)$$

The ionization potential and electron affinity of expressions (69) can be used to derive the *chemical* electronegativity and hardness, through equations (11) and (13), respectively, performing the finite integrals of the absolute electronegativity (10):

$$\chi_C = \frac{IP + EA}{2} = \frac{1}{2} \int_{N_v-1}^{N_v+1} \chi dN, \qquad (70)$$

$$\eta_C \equiv \frac{IP - EA}{2} = \chi_C - \int_{N_v}^{N_v+1} \chi dN + C_A \qquad (71)$$

From equations (70) and (71) one can observe that the chemical electronegativity writes as simple average of the absolute one over the exchanged charge path (16), while chemical hardness depends both on chemical electronegativity and chemical action. However, there is clear that the knowledge of the absolute electronegativity (10) is compulsory in order to have as an output the associate chemical electronegativity and hardness.

In this respect, recalling the nature of the thermodynamical potential of electronegativity via its direct relation with chemical potential, one can consider the representation $\chi = \chi[N, V(x)]$ also for the absolute electronegativity by means of the first order N- and $V(x)$- expansion:

$$d\chi = \left(\frac{\partial \chi}{\partial N}\right)_{V(x)} dN - \int \left(\frac{\partial \rho(x)}{\partial N}\right)_{V(x)} dV(x) dx, \qquad (72)$$

where accounted of equation (67).

Nevertheless, more meaningful, equation (72) can be fashioned as:

$$d\chi = -\frac{1}{S} dN - \int \frac{s(x)}{S} dV(x) dx \qquad (73)$$

since the local softness,

$$s(x) = -\left(\frac{\partial \rho(x)}{\partial \chi}\right)_{V(x)}, \qquad (74)$$

and the global one (24),

$$S = \int s(x) dx = -\left(\frac{\partial N}{\partial \chi_D}\right)_{V(x)}, \qquad (75)$$

are respectively introduced, counting on the fundamental *DFT* expression (28) and of the absolute hardness relation with absolute electronegativity (12).

Now, in terms of local and global softness, the differential electronegativity equation (73) can be integrated to give:

$$\chi = -\int_0^{N_V} \frac{1}{S} dN - \int \frac{s(x)}{S} \left(\int_0^{V(x)} dV \right) dx = -\int_0^{N_V} \frac{1}{S} dN - \frac{1}{S} \int s(x)V(x) dx \tag{76}$$

From the hierarchically point of view, the absolute electronegativity (76) can be considerate as the starting point in deriving the chemical electronegativity and hardness – see equations (70) and (71), respectively. However, the dependence is on the softness rather than on the total energy.

To summarize, all working absolute and chemical levels of electronegativity and hardness are collected in the table 2.

Table 2. Different electronegativity (left column) and hardness (right column) in the absolute (first two rows) and chemical (last two rows) formulations relating the local and global softness contributions.

χ	η
$\chi = -\int_0^{N_V} \frac{1}{S} dN - \frac{1}{S} \int s(x)V(x)dx$	$\eta^{\chi} = -\frac{1}{2}\left(\frac{\partial \chi}{\partial N}\right)_V$
	$\eta^S = \frac{1}{2S}$
$\chi_C = \frac{1}{2} \int_{N_v-1}^{N_v+1} \chi dN$	$\eta_C^{\chi} = -\frac{1}{2}\left(\frac{\partial \chi_C}{\partial N}\right)_V$
	$\eta_C^{CA} = \chi_C - \int_{N_v}^{N_v+1} \chi dN + C_A$

Nevertheless, since the global softness (24) is quite insensitive to the number of electrons one can further assume that the present softness picture holds for every value of the number of valence electrons, starting at zero and culminating in the desired value.

From table 2 there appears that a proper softness formulation should be further considerate in order to obtain the electronegativity and hardness density functional formulations.

Fortunately, such a density functional softness theory exists [5], and it is based on the softness kernel local-nonlocal picture [109]:

$$s(x,x') = L(x')\delta(x-x') + \rho(x)\rho(x'), \qquad (77)$$

whose the local contribution is regulated by the response function:

$$L(x) = -\frac{\nabla\rho(x)\cdot\nabla V(x)}{|\nabla V(x)|^2} \qquad (78)$$

that modulates the delta-Dirac function $\delta(x'-x)$.

Worth noting that the above approximate softness kernel formula (77) is sustained by three quantum mechanical constraints: the translational invariance condition [129], the Hellmann-Feynman theorem [126], and the normalization of the linear response function [110], are providing therefore sufficient rigorous framework for further analytical developments.

This way, the hierarchy of the softness density functional picture recovers the local softness form from the kernel one (77) via simple integration:

$$s(x) = \int s(x,x')dx' = L(x) + N_v\rho(x) \stackrel{(78)}{=} -\frac{\nabla\rho(x)}{\nabla V(x)} + N_v\rho(x), \qquad (79)$$

while global softness will acquire explicit density functional expression through further integration of local softness (79) over the remaining coordinate:

$$S = \int s(x)dx = \int L(x)dx + N_v^2 \stackrel{(78)}{=} N_v^2 - \int\frac{\nabla\rho(x)}{\nabla V(x)}dx \qquad (80)$$

Finally, with local and global softness expressions, (79) and (80), respectively, the absolute and chemical softness-related electronegativity and hardness density functionals of table 2 become analytic density indices (in atomic units). Thus, the whole class of the global reactivity descriptors, here presented, can be expressed within a consistently $[\rho(x), V(x)]$ – DFT based context, at various degrees of local and nonlocal electronic effects and comprehension, as will be in next revealed.

4.2. LOCAL LIMITED SOFTNESS EFFECTS

The exclusive local softness formulation for electronegativity and hardness is based on the expressions (79) and (80) in which the non-local effects are totally neglected. This approach, were recently employed to analyze the HSAB effects by assuming the electronic effects associated with neglecting of the exchange-correlation terms far away of nuclei, where the electronic density fall-off [171]. In this context the working softness quantities are approximated as:

$$S \cong \int L(x)dx \equiv a,$$
$$s(x) \cong L(x), \qquad (81)$$

where, practically, the analytical conditions $\rho(x) \to 0$ or $\rho(x') \to 0$ were stated.

With relations (81) the associate analytical N-dependent expressions for absolute and chemical electronegativity and hardness results by solving the table 2. For instance, the absolute electronegativity comes out by integration in (76):

$$\chi = -\frac{N_v + b}{a}, \qquad (82)$$

where another density functional response index was introduced:

$$b \equiv \int L(x)V(x)dx \qquad (83)$$

Here, worth remarking that the various response density functionals that emerged from the softness kernel model (79), like a and b of (81) and (83), respectively, are formally assumed to be "independent" of the number of electrons. These assumptions cannot be rigorously justified but on the *post-facto* ground, being applied due to the flexibility in further mathematically manipulations that they induce [75].

With absolute electronegativity (82) the chemical electronegativity (70) can be immediately inferred, with the nice result:

$$\chi_C = \frac{1}{2}\int_{N_v-1}^{N_v+1}\chi dN = -\frac{N_v+b}{a} = \chi \qquad (84)$$

However, the identity (84) gives us an important fact, namely that the absolute electronegativity (the minus of chemical potential) and the chemical (or integral) electronegativity are rigorously equal when the non-local effects are neglected.

Table 3. Limited electronegativity (left column) and hardness (right column) in the absolute (first two rows) and chemical (last two rows) formulations, abstracted from table 2, when the local softness approximation (81) is employed

χ_L	η_L
$\chi = -\dfrac{N_v+b}{a}$	$\eta^\chi = \dfrac{1}{2a}$
	$\eta^S = \dfrac{1}{2a}$
$\chi_C = -\dfrac{N_v+b}{a}$	$\eta_C^\chi = \dfrac{1}{2a}$
	$\eta_C^{CA} = \dfrac{1}{2a}+C_A$

The rest of quantities of table 2 are reported in table 3 with their analytical form for the case of accounting for the limited effects of local softness only. Hereafter the "limited" nomination will be the characteristic appellative of this case, instead of "local" one, avoiding to induce the paradoxically idea of local-global indices of reactivity.

Nevertheless, from table 3 one can clearly seen that the exclusive local effects consists in a drastic unification of absolute and chemical electronegativity and hardness formulations through the simple density functionals, (81) and (83).

The unified expressions in this case will be called as (softness) *limited electronegativity and hardness*,

$$\chi_L = -\frac{N_v + b}{a}, \qquad (85)$$

$$\eta_L = \frac{1}{2a}, \qquad (86)$$

respectively, being different from that one which explicitly contains the chemical action correction, called hereafter (softness) *limited chemical action hardness*:

$$\eta_L^{CA} = \eta_L + C_A \qquad (87)$$

Note that the chemical action correction appears only to the chemical hardness, not to electronegativity, emphasizing on the regulating role that chemical action carries in the chemical bonding scenario, see table 1 and the accompanied discussion, where it is added to the chemical hardness as the averaged quantum fluctuation targeting the final stabilization of bond through the maximum hardness realization. Thus, this "plus value" chemical action contributes to stabilize the maximum chemical hardness at the end of bonding process.

Another observation regards the cases where the chemical action does not appear explicitly. Still, its implicit influence is sensitively induced by means of the $[\rho(x), V(x)]$ combinations that appear also in the definitions of a and b functionals, (81) and (83), via the local response function (78). This can be better seen if one rewrites the function (78) as

$$L(x) = \frac{1}{2} \frac{\nabla^2(\rho(x)V(x)) - \rho(x)\nabla^2 V(x) - V(x)\nabla^2 \rho(x)}{|\nabla V(x)|^2}, \qquad (88)$$

where, we can recognize that

$$\rho(x)V(x) = |\nabla C_A| \qquad (89)$$

according with the basic chemical action definition, (35) or (48).

More, as we already learn from the chemical action interpretation, the definite integrals of response indices (81) and (83) can be decomposed on the appropriate sum of orbitals, as was done in (51), to further unify the orbital with the global nature of the computed "local" electronegativity and hardness density functionals (85)-(87).

4.3. [N] - REPRESENTABLE NONLOCAL SOFTNESS EFFECTS

The next interesting case consists in considering as the starting point the full softness density functional formulation form (80) when, from the basic variation of the absolute electronegativity $[N, V(x)]$ representation (72), only the charge transfer is considered, within a bonding picture based on vertical potential interaction, thus neglecting the relaxation effects in external potential rearrangements.

This approach corresponds in the bonding scenario of table 1 with the situation in which the quantum effects associated with the inequality electronegativity principle (EI) are neglected. It however leads with an implicit fulfillment of the hardness global equilibrium and an instantaneous HSAB and MH applied principles.

Analytically, in this case, the absolute electronegativity of the table 2 restricts itself to the expression:

$$\chi = -\int_0^{N_v} \frac{1}{S} dN = -\frac{1}{\sqrt{a}} \arctan\left(\frac{N_v}{\sqrt{a}}\right), \qquad (90)$$

while the absolute and related softness hardness takes the unified form:

$$\eta^{\chi} = \eta^{S} = \frac{1}{2(a+N_v^2)} \qquad (91)$$

Passing to the chemical variant of the concerned reactivity indices, the chemical electronegativity and hardness are respectively obtained:

$$\chi_C = \frac{N_v-1}{2\sqrt{a}} \arctan\left(\frac{N_v-1}{\sqrt{a}}\right) - \frac{N_v+1}{2\sqrt{a}} \arctan\left(\frac{N_v+1}{\sqrt{a}}\right) - \frac{1}{4}\ln\left[\frac{a+(N_v-1)^2}{a+(N_v+1)^2}\right], \quad (92)$$

$$\eta_C^\chi = \frac{1}{4\sqrt{a}} \left[\arctan\left(\frac{N_v+1}{\sqrt{a}}\right) - \arctan\left(\frac{N_v-1}{\sqrt{a}}\right) \right], \qquad (93)$$

at the same time as the chemical action adds directly to the chemical hardness thus furnishing the chemical action hardness index,

$$\eta_C^{CA} = \frac{N_v-1}{2\sqrt{a}} \arctan\left(\frac{N_v-1}{\sqrt{a}}\right) - \frac{N_v}{\sqrt{a}} \arctan\left(\frac{N_v}{\sqrt{a}}\right) + \frac{N_v+1}{2\sqrt{a}} \arctan\left(\frac{N_v+1}{\sqrt{a}}\right)$$
$$+ \frac{1}{4} \ln \left\{ \frac{(a+N_v^2)^2}{[a+(N_v+1)^2][a+(N_v-1)^2]} \right\} + C_A, \qquad (94)$$

this way emphasizing, as previously, on the major role of the chemical action in assuring the maximum hardness values and variational principle.

Aiming to deal with unified versions of the above (90)-(94) electronegativity and hardness density formulations worth searching for appropriate limits in which they converge. In this respect, unlike the previous case, the nonlocal effects are explored here, that is to apply the equivalent limits:

$$L(x) \xrightarrow{(78)} 0 \Leftrightarrow \nabla \rho(x) \xrightarrow{(81)} 0 \Leftrightarrow a \to 0, \qquad (95)$$

together with the statistical one:

$$N_v \to \infty, \qquad (96)$$

in various orders of expansions. Only in these conditions the relations (90)-(94) find the proper unified formulations, as presented in the table 4.

A special remark deserves here. Restricting now to the non-local effects means counting exclusively on exchange and correlations between electrons thus giving them the priority role in deciding the bonding and reactivity. On the other hand, the statistical limit (96) requires that electrons correlate in such manner to permit an infinite collection of them on the same state; this is nothing than the *bosonic* state of matter. More interesting, the limit (96) acquires in the same time also the significance for the pure statistical fermionic samples, without exchange and correlation, being this the limit in which the Thomas-Fermi model becomes exact, as the extreme case of the density functional theory [7].

Table 4. First bosonic electronegativity (left column) and hardness (right column) in the absolute (first two rows) and chemical (last two rows) formulations, abstracted from table 2, when the nonlocal softness limit ($\nabla\rho(x) \to 0$ or $a \to 0$) together with different orders of statistical limit ($N_v \to \infty$) are employed on relations (90)-(94)

$\chi_B^{[I]}$	$\eta_B^{[I]}$
$\chi = \dfrac{1}{N_v} - \dfrac{\pi}{2\sqrt{a}}$	$\eta^\chi = \dfrac{1}{2N_v^2}$
	$\eta^S = \dfrac{1}{2N_v^2}$
$\chi_C = \dfrac{1}{N_v} - \dfrac{\pi}{2\sqrt{a}}$	$\eta_C^\chi = \dfrac{1}{2N_v^2}$
	$\eta_C^{CA} = \dfrac{1}{2N_v^2} + C_A$

From these last two issues results that a special boson-fermionic mixture is formed when absolute and chemical electronegativity and hardness unify as in table 4; nevertheless, to underline the special feature of this case, based on the joint limits (95) and (96), they are called as *first bosonic density functionals of electronegativity and hardness*, respectively:

$$\chi_B^{[I]} = \frac{1}{N_v} - \frac{\pi}{2\sqrt{a}}, \tag{97}$$

$$\eta_B^{[I]} = \frac{1}{2N_v^2} \tag{98}$$

Of course, as observed form table 4, the special behavior of the chemical action is manifested as the "plus value" also to the first bosonic chemical hardness as well,

$$\eta_B^{CA[I]} = \eta_B^{[I]} + C_A,\qquad(99)$$

thus directly contributing to the regulation of the maximum chemical hardness by the resulted *first bosonic chemical action hardness*.

4.4. [N, V] - REPRESENTABLE NONLOCAL SOFTNESS EFFECTS

In the same manner as before, the full $[N, V(x)]$ representation of absolute electronegativity equation (72) can be considered to the limiting (95) and (96) nonlocal effects of softness (80). Firstly, making use of the entirely softness picture, through relations (79) and (80), one can straightly get the analytical counterpart of the expression (76):

$$\chi = -\frac{1}{\sqrt{a}}\arctan\left(\frac{N_v}{\sqrt{a}}\right) - \frac{b}{a+N_v^2} - N_v C_A \frac{1}{a+N_v^2},\qquad(100)$$

with the help of which the corresponding absolute hardness comes out by applying the derivative definition (12) with density functional result:

$$\eta^\chi = \frac{1}{2(a+N_v^2)} + \frac{(a-N_v^2)C_A - 2N_v b}{2(a+N_v^2)^2},\qquad(101)$$

while the softness related one maintains the previous dependence (91).

With expressions (100) and (101) the rest of the indices from the table 2 take now their specific analytical density functional forms:

$$\chi_C = \frac{b+N_v-1}{2\sqrt{a}}\arctan\left(\frac{N_v-1}{\sqrt{a}}\right) - \frac{b+N_v+1}{2\sqrt{a}}\arctan\left(\frac{N_v+1}{\sqrt{a}}\right)$$
$$+\frac{C_A-1}{4}\ln\left[\frac{a+(N_v-1)^2}{a+(N_v+1)^2}\right],\qquad(102)$$

$$\eta_C^\chi = \frac{N_v^4 + 2N_v^2(a-1) + (a+1)^2}{4\sqrt{a}\left[a+(N_v-1)^2\right]\left[a+(N_v+1)^2\right]}\left[\arctan\left(\frac{N_v+1}{\sqrt{a}}\right) - \arctan\left(\frac{N_v-1}{\sqrt{a}}\right)\right]$$
$$+ \frac{(1+a-N_v^2)C_A - 2N_v b}{2\left[a+(N_v-1)^2\right]\left[a+(N_v+1)^2\right]}, \qquad (103)$$

$$\chi_C^{CA} = \frac{b+N_v-1}{2\sqrt{a}}\arctan\left(\frac{N_v-1}{\sqrt{a}}\right) - \frac{b+N_v}{\sqrt{a}}\arctan\left(\frac{N_v}{\sqrt{a}}\right) + \frac{b+N_v+1}{2\sqrt{a}}\arctan\left(\frac{N_v+1}{\sqrt{a}}\right)$$
$$+ \frac{1-C_A}{4}\ln\left\{\frac{(a+N_v^2)^2}{\left[a+(N_v+1)^2\right]\left[a+(N_v-1)^2\right]}\right\} + C_A \qquad (104)$$

There is obvious that, in this case, more than in the previous one, there appears that none of the above analytical indices display identical working formulations. To achieve the unification goal the limits (95) and (96) are again considered at different orders so that to release the convergent density functionals versions of absolute and chemical electronegativity and hardness. The unified results are presented in the table 5. However, from it, there outcomes that whereas electronegativity becomes equal expressed in both of its natural versions, as the so called *second bosonic electronegativity*:

$$\chi_B^{[II]} = \left(\frac{1}{N_v} - \frac{\pi}{2\sqrt{a}}\right) - \frac{1}{N_v}\left(C_A + \frac{b}{N_v}\right), \qquad (105)$$

the respective hardness expressions still present some inherent differences even in the conditions of the unified frame of electronegativity.

As such, since the softness-related hardness preserves the previous first bosonic limit (98) will be not repeated here as a distinct *second bosonic hardness* that, however, will be attributed to that abstracted from the unified absolute with chemical electronegativity, namely:

$$\eta_B^{[II]} = \frac{1-C_A}{2N_v^2} - \frac{b}{N_v^3} \qquad (106)$$

In any case, as already noted, the second bosonic chemical action hardness is obtained from the second bosonic hardness (106) to which the chemical action adds,

$$\eta_B^{CA[II]} = \eta_B^{[II]} + C_A, \tag{107}$$

to characterize the averaged chemical quantum effects compressed in maximum hardness stabilization principle.

Table 5. Second bosonic electronegativity (left column) and hardness (right column) in the absolute (first two rows) and chemical (last two rows) formulations, abstracted from table 2, when the nonlocal softness limit ($\nabla \rho(x) \to 0$ or $a \to 0$) together with different orders of statistical limit ($N_v \to \infty$) are employed on relations (91) and (100)-(104)

$\chi_B^{[II]}$	$\eta_B^{[II]}$
$\chi = \left(\dfrac{1}{N_v} - \dfrac{\pi}{2\sqrt{a}}\right) - \dfrac{1}{N_v}\left(C_A + \dfrac{b}{N_v}\right)$	$\eta^\chi = \dfrac{1-C_A}{2N_v^2} - \dfrac{b}{N_v^3}$
	$\eta^S = \dfrac{1}{2N_v^2}$
$\chi_C = \left(\dfrac{1}{N_v} - \dfrac{\pi}{2\sqrt{a}}\right) - \dfrac{1}{N_v}\left(C_A + \dfrac{b}{N_v}\right)$	$\eta_C^\chi = \dfrac{1-C_A}{2N_v^2} - \dfrac{b}{N_v^3}$
	$\eta_C^{CA} = \dfrac{1-C_A}{2N_v^2} - \dfrac{b}{N_v^3} + C_A$

However, all presented limited and statistical or bosonic density functional electronegativity and hardness will be next evaluated at the atomic level and then implemented at the molecular level providing therefore the suitable platform for practicing the reactivity principles of chemical bonding of table 1.

Chapter 5

5. ATOMIC ELECTRONEGATIVITY AND HARDNESS

Going to apply the introduced unified density functionals of electronegativity and hardness the computation of associated atomic scales stands as the natural next step. Yet, before exposing the specific procedure of computation and discuss the results let's remember the main analytical approximation used in the present endeavor so far. They consist in:

- adopting the absolute definitions for electronegativity and hardness through the basic first and second total energy derivatives - see equations (10) and (12), respectively;
- adopting the chemical definitions for electronegativity and hardness through semi-sum and semi-difference of ionization potential and electron affinity – see equations (11) and (13), respectively, with the reactivity path integral realizations of table 2;
- considering the first order truncation of the $[N, V(x)]$ – expansion of absolute electronegativity – see equation (72), based on the $[N, V(x)]$ – representability of the total energy (40);
- restricting the effective number of electrons to those of the valence shell, N_v.

However, the remaining conditions to be explicit implemented in the present approach, for consistency, regard the N - dependency of density functionals a, b, and C_A of equations (81), (83), and (35), respectively, together with the explicit

$V(x)$ – dependency of $\rho(x)$, as required by the basic context of the Hohenberg-Kohn theorems of *DFT*.

5.1. COMPUTATIONAL METHOD

The atomic case will be considered by employing the simple yet meaningful Slater's asymptotic large distance picture, for valence shells, for a hydrogenic-like wave function in the central field of an *effective nuclear charge* Z^*,

$$V^*(r) = -\frac{Z^*}{r}, \qquad (108)$$

due to the screening effects of the inner electrons [172].

As consequence, the desired effective radial density at the valence shell level, when N_v electrons are involved, takes the working form [75]:

$$\rho_n^*(r) = N_v \frac{(2\xi^*)^{2n+1}}{(2n)!} r^{2n} \exp(-2\xi^* r), \qquad (109)$$

being characterized, beside the principal quantum number n, by the *orbital exponent* ξ^*. The values of Z^* and ξ^* are computed upon specific rules constructed so that the associate energy levels to check fairly with experiment [172-176].

In these conditions, the chemical action density functional (35) is firstly written as:

$$C_A(N_v, n, \xi^*, Z^*) = 4\pi \int_0^\infty \rho_n^*(r) V^*(r) r^2 dr$$

$$= -4\pi Z^* N_v \frac{(2\xi^*)^{2n+1}}{(2n)!} \int_0^\infty r^{2n+1} \exp(-2\xi^* r) dr \qquad (110)$$

Still, in order to evaluate the above integral the valence shell properties are employed by performing the saddle point approximation around the most probable radius of atom r_0, from the perspective of the quantity to be evaluated.

This way, the integrals of the type

$$I(\alpha,q) = \int_0^\infty r^\alpha \exp(-qr)dr$$

$$= \int_0^\infty \exp[\alpha \ln r - qr]dr = \int_0^\infty \exp\left[\alpha\left(\ln r - q\frac{r}{\alpha}\right)\right]dr$$

$$\equiv \int_0^\infty \exp[\alpha f(r)]dr, \; \alpha > 0 \qquad (111)$$

are approximated as [75]:

$$I(\alpha,q) \cong \exp[\alpha f(r_0)]\sqrt{-\frac{2\pi}{\alpha f''(r_0)}}$$

$$= \exp\left[\alpha\left(\ln\frac{\alpha}{q}-1\right)\right]\sqrt{\frac{2\pi\alpha}{q^2}} = \sqrt{2\pi}\exp(-\alpha)\frac{\alpha^{\alpha+\frac{1}{2}}}{q^{\alpha+1}}, \qquad (112)$$

computed at the saddle point $r_0 = \alpha/q$ where the phase function $f(r) = \ln r - qr/\alpha$ of (111) fulfills the stationary condition $f'(r) = 0$.

With the recipe (112), the chemical action expression (110) takes the actual working form:

$$C_A(N_v,n,\xi^*,Z^*) = -2\sqrt{2}\pi^{3/2}\frac{N_v Z^*}{\xi^*}\frac{(2n+1)^{2n+3/2}}{(2n)!}\exp[-(2n+1)] \qquad (113)$$

In the same manner, the response function (78) is primarily written:

$$L(r, N_v, n, \xi^*, Z^*) = \frac{[\nabla \rho_n^*(r)][-\nabla V^*(r)]}{[-\nabla V^*(r)]^2}$$

$$= 2\frac{N_v}{Z^*}\frac{(2\xi^*)^{2n+1}}{(2n)!}\left(\xi^* r^{2n+2} - nr^{2n+1}\right)\exp(-2\xi^* r), \quad (114)$$

to be then used in evaluation of the response density functionals (81) and (83), respectively:

$$a(N_v, n, \xi^*, Z^*) = 4\pi \int_0^\infty L(r, N_v, n, \xi^*, Z^*) r^2 dr$$

$$= 8\pi \frac{N_v}{Z^*}\frac{(2\xi^*)^{2n+1}}{(2n)!}\left[\xi^* \int_0^\infty r^{2n+4} \exp(-2\xi^* r) dr - n\int_0^\infty r^{2n+3} \exp(-2\xi^* r) dr\right]$$

$$= \pi^{3/2} \frac{N_v}{Z^* \xi^{*3}(2n)!}\left(\frac{2}{e}\right)^{2(n+2)}\left[(n+2)^{2n+9/2} - en\left(n+\frac{3}{2}\right)^{2n+7/2}\right], \quad (115)$$

$$b(N_v, n, \xi^*) = 4\pi \int_0^\infty L(r, N_v, n, \xi^*, Z^*) V^*(r) r^2 dr$$

$$= 8\pi N_v \frac{(2\xi^*)^{2n+1}}{(2n)!}\left[\xi^* \int_0^\infty r^{2n+3} \exp(-2\xi^* r) dr - n\int_0^\infty r^{2n+2} \exp(-2\xi^* r) dr\right]$$

$$= \pi^{3/2} \frac{N_v}{\xi^{*2}(2n)!}\frac{4^{n+2}}{\exp(2n+3)}\left[e n(n+1)^{2n+5/2} - \left(n+\frac{3}{2}\right)^{2n+7/2}\right] \quad (116)$$

With expressions (113), (115), and (116) the systematic (softness) limited-, first bosonic-, second bosonic- and chemical action- related electronegativity and hardness, see the sets of equations (85)-(87), (97)-(99), (105)-(107), and (57)-(60), respectively, can be computed since their particular formulations are employed upon a certain atomic system with the relevant parameters, as displayed in the table 6.

Table 6. The periodic input parameters used in the actual study: the total number of s+p electrons, N_v, the principal quantum number n, the orbital exponent ξ^* and the effective charge Z^*, calculated upon Slater method [172], for the valence shells of the ordinary elements.

H	He	Legend: Symbol of Element
1	2	Number of $s + p$ valence electrons: N_v
1	1	Valence principal quantum number: n
1	1.7	Valence orbital exponent: ξ^*
1	1.7	Valence effective charge: Z^*

Li	Be											B	C	N	O	F	Ne
1	2											3	4	5	6	7	8
2	2											2	2	2	2	2	2
0.65	0.98											1.3	1.63	1.95	2.28	2.6	2.93
1.30	1.95											2.60	3.25	3.90	4.55	5.2	5.85
Na	Mg											Al	Si	P	S	Cl	Ar
1	2											3	4	5	6	7	8
3	3											3	3	3	3	3	3
0.73	0.95											1.17	1.39	1.6	1.8	2.03	2.25
2.20	2.85											3.50	4.15	4.80	5.45	6.10	6.75
K	Ca	Sc	Ti	V	Cr	Mn	Fe	Co	Ni	Cu	Zn	Ga	Ge	As	Se	Br	Kr
1	2	2	2	2	2	2	2	2	2	2	2	3	4	5	6	7	8
4	4	4	4	4	4	4	4	4	4	4	4	4	4	4	4	4	4
0.59	0.77	0.81	0.85	0.89	0.93	0.97	1.01	1.05	1.09	1.14	1.18	1.35	1.53	1.70	1.88	2.05	2.23
2.20	2.85	3.00	3.15	3.30	3.45	3.60	3.75	3.90	4.05	4.20	4.35	5.00	5.65	6.30	6.95	7.60	8.25
Rb	Sr	Y	Zr	Nb	Mo	Tc	Ru	Rh	Pd	Ag	Cd	In	Sn	Sb	Te	I	Xe
1	2	2	2	2	2	2	2	2	2	2	2	3	4	5	6	7	8
5	5	5	5	5	5	5	5	5	5	5	5	5	5	5	5	5	5
0.55	0.71	0.75	0.79	0.83	0.86	0.9	0.94	0.98	1.01	1.05	1.09	1.25	1.41	1.58	1.74	1.9	2.06
2.20	2.85	3.00	3.15	3.30	3.45	3.60	3.75	3.90	4.05	4.20	4.35	5.00	5.65	6.30	6.95	7.60	8.25

Certainly, by the present way of computation another set of approximations have been assumed, this time at the numerical atomic level, as follows:

- assuming one-electron picture under the hydrogenic radial density function formulation, see equation (109);
- setting the largest quantum number of orbital angular momentum, $l = n-1$, in the general Laguerre polynomials, such that to deal only with one maxima – or 0 nodes – in the working density function (109);

- employing the asymptotic Slater approach, properly for electronegativity and hardness modelling of bonding where the interactions are envisaged from the valence shells, so using the effective nuclear charge and orbital exponent, Z^* and ξ^*, respectively, with the individual values for atoms given in table 6.
- transforming the working electronic density function by multiplying it with the number of valence electrons, see equation (109), so fulfill the *DFT* basic constraint (28) and becoming compatible with the many-electronic definition of electronegativity and hardness through the table 2;
- applying the saddle point approximation to evaluate the involved integrals in computing chemical action (35) and the response functionals (81) and (83), providing the working valence formulas (113), (115) and (116), respectively.

Table 7. Calibration coefficients of the electronegativity χ and hardness η for the considered chemical action -, (softness) limited-, first- and second bosonic- related electronegativity and hardness, see the sets of equations (59)-(60), (85)-(87), (97)-(99), and (105)-(107), respectively, such way their values for atomic H to recover their counterpart finite differences, computed upon the IP and EA formulas, (11) and (13), having the 7.18 and 6.45 eV (electron-volts), respectively.

Source	χ	η
[chemical action]	$\chi_{CA} \times 27.21 \times (19.6712)$	$\eta_{CA} \times 27.21 \times (-17.6712)$
[softness limited]	$\chi_L \times 27.21 \times (0.888776)$	$\eta_L \times 27.21 \times (176.509)$
		$\eta_L^{CA} \times 27.21 \times (-0.0129295)$
[first bosonic]	$\chi_B^{[I]} \times 27.21 \times (0.287259)$	$\eta_B^{[I]} \times 27.21 \times (0.47409)$
		$\eta_B^{CA[I]} \times 27.21 \times (-0.013291)$
[second bosonic]	$\chi_B^{[II]} \times 27.21 \times (0.00201752)$	$\eta_B^{[II]} \times 27.21 \times (0.00195574)$
		$\eta_B^{CA[II]} \times 27.21 \times (0.00230432)$

Finally, before computing the atomic scales, the calibration procedure has to be also performed respecting the experimental-based electronegativity and hardness values for the H atom, 7.18 and 6.45eV (electron-volts), by means of the measured values of IP_H and EA_H, 13.62 and 0.73 eV, in the formulas (11) and (13), respectively, for each considered electronegativity and hardness density functional introduced in this study; the results are presented in table 7.

5.2. RESULTS AND DISCUSSIONS

Originally, in 1869, the periodic system was the fruit of the Mendeleyev experimental observation that the physical and chemical properties of elementary chemical substances vary as periodic functions with their atomic weight. Fortunately, he arranged his chart so that also the atomic number Z had played an important role, being later confirmed by the X-ray studies of Moseley in 1913. More, with the advent of quantum mechanical description of atoms and molecules there was clear that the elements are arranged in a bi-dimensional space according with their number of shells and their occupancy (or valence) down groups and along periods, respectively.

However, despite some objections in the sense that this so called *aufbau principle* is merely based on experiment and do not corresponds with an inner structural quantum information [15], there is clear that as far as electronic occupancy in atoms takes a gradual character the exchange-correlation effects are included in all functions that combines the periodic quantum number that characterize a certain element.

There was therefore a challenge to find the proper functions that making that job provide the true theoretical, or intrinsic, character of the quantum constitution of atoms, without invoking necessarily the electronic spin effects. Consequently, the third dimension of Periodic System is to be associated with fulfillment or proving the aufbau principle.

In this respect, Leland Allen was first to recognize that the electronegativity concept and its scale can indeed furnish the appropriate function to reproduce the observed periodicities of experimental quantities, see [13] and the references therein, as the ionization potentials and electronic affinities, for instance.

Here, one note is nevertheless useful for what next and to justify to some extent why we have so many atomic scales for the same quantity, the electronegativity, at our disposal.

We concentrate only on the basic chemical definition (9) to sustain this issue. As we see, the chemical electronegativity implies the knowledge or evaluation of two quantities, *IP* and *EA*. However, they are conceptually distinct since the first is manifested only when additional energy is furnished to the electronic system while the second one unfolds as the inner propensity for electronic attraction without any perturbation on the system.

In short, it seems that the *IP* is more environmental quantity while *EA* stands as the more intrinsic one. Phenomenologically, this situation is in full compatible with the actual interpretation of electronegativity as the bridge between the virtual and manifested natures of chemical bonding, here combined through *EA* and *IP* propensities, respectively.

The lesson is clear: as quantity caries more virtual character the lesser experimentally accessible is. Indeed, the more structural characteristic *EA* gives much more experimentally difficulties in order to measure it; the situation is reverse for *IP* nature and assaying [174, 175]. The same characteristics can be transferred to the present day concept of hardness: being associated with the second order effects, i.e. the stabilization of the electronegativity variation, it characterizes even more specific, or isolated, or virtual, character of the concerned electronic system. Only for this reason there is not justification to reject or to deny the χ and η role in chemical reactivity [15] as they fully drive the chemical potential and force in bonding.

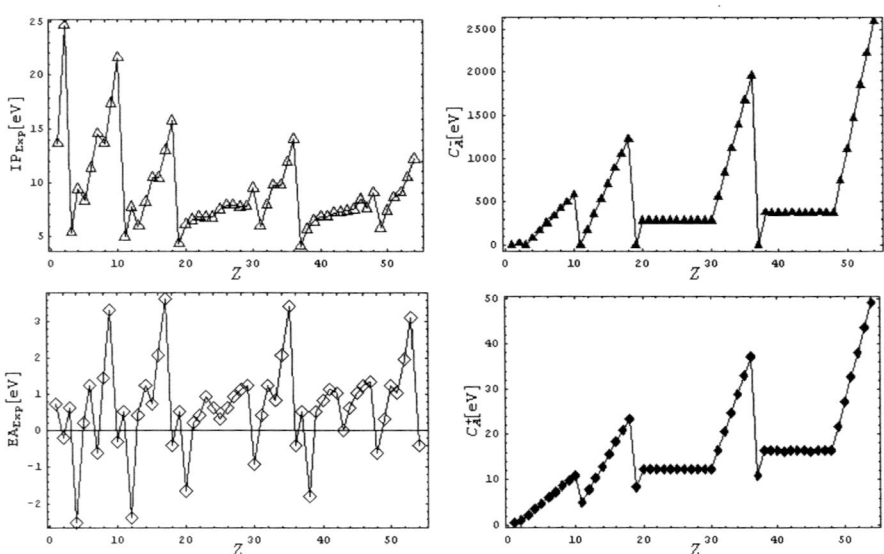

Figure 3. From up to down: experimental first ionization potentials and the first electron affinities (on the left column), together with the corresponded HOMO- and LUMO- chemical actions (on right column), from table 8, respectively, for the ordinary elements.

Atomic Electronegativity and Hardness 53

Table 8. Experimental first ionization potentials, first electron affinities [162], and the corresponded HOMO- and LUMO- chemical actions calculated upon equations (57) and (58) by specializing the relation (113), respectively, in eV (electron-volts), for the ordinary elements.

		H	He	*Legend*: Symbol of Element													
		13.62	24.65	Experimental Ionization Potential: IP_{Exp}													
		0	24.65	HOMO Chemical Action: C_A^-													
		0.73	-0.21	Experimental Electron Affinity: EA_{Exp}													
		0.73	1.1	LUMO Chemical Action: C_A^+													
Li	Be											B	C	N	O	F	Ne
5.41	9.36											8.32	11.34	14.56	13.62	17.47	21.63
0	82.65											166.16	248.47	332.31	414.48	489.47	580.56
0.62	-2.5											0.21	1.25	-0.62	1.46	3.33	-0.31
2.46	3.67											4.92	6.13	7.38	8.59	9.84	11.05
Na	Mg											Al	Si	P	S	Cl	Ar
5.02	7.7											6.03	8.22	10.5	10.4	13	15.81
0	175.3											349.59	523.36	701.18	884.59	1053.5	1227.06
0.52	-2.39											0.42	1.25	0.73	2.08	3.64	-0.42
5.21	7.79											10.35	12.92	15.57	18.34	20.8	23.36
K	Ca	Sc	Ti	V	Cr	Mn	Fe	Co	Ni	Cu	Zn	Ga	Ge	As	Se	Br	Kr
4.37	6.14	6.55	6.86	6.76	6.76	7.49	7.90	7.90	7.7	7.8	9.46	6.03	7.90	9.78	9.78	11.86	14.04
0	278.8	278.98	279.15	279.29	279.43	279.56	279.67	279.78	279.88	277.51	277.68	557.96	834.48	1116.58	1392.31	1675.52	1950.68
0.52	-1.66	0.21	0.42	0.94	0.62	0.31	0.62	0.94	1.14	1.25	-0.94	0.42	1.25	0.83	2.08	3.43	-0.42
8.32	12.38	12.39	12.4	12.41	12.42	12.42	12.43	12.33	12.34	1e-53	20.59	24.8	28.86	33.08	37.14		
Rb	Sr	Y	Zr	Nb	Mo	Tc	Ru	Rh	Pd	Ag	Cd	In	Sn	Sb	Te	I	Xe
4.16	5.72	6.45	6.86	6.86	7.18	7.28	7.38	7.49	8.42	7.59	9.05	5.82	7.38	8.63	9.05	10.5	12.17
0	370.17	368.87	367.71	366.65	369.95	368.87	367.89	366.99	369.79	368.87	737.75	1108.59	1470.83	1841.73	2213.25	2585.26	
0.52	-1.77	0.52	0.83	1.14	1.04	0.00	0.62	1.04	1.25	1.35	-0.62	0.31	1.25	1.04	1.98	3.12	-0.42
10.92	16.44	16.39	16.33	16.29	16.43	16.39	16.34	16.3	16.43	16.39	16.35	21.85	27.36	32.67	38.18	43.7	49.22

Table 9. The finite difference (or experimental related) chemical electronegativity and chemical hardness calculated with the help of IP and EA of table 8 upon equations (11) and (13), together with the chemical action -electronegativity and -hardness computed from equations (59) and (60), respectively, in eV (electron-volts), for the ordinary elements.

		H	He	*Legend*: Symbol of Element													
		7.18	12.27	Finite Difference Electronegativity: χ_{FD}													
		7.18	253.22	Chemical Action Electronegativity: χ_{CA}													
		6.45	12.48	Finite Difference Hardness: η_{FD}													
		-6.45	208.1	Chemical Action Hardness: $-\eta_{CA}$													
Li	Be											B	C	N	O	F	Ne
3.02	3.43											4.26	6.24	6.97	7.59	10.4	10.71
24.2	849.07											1682.7	2504.2	3341.1	4161.2	4999.6	5818.8
4.39	5.93											4.06	4.99	7.59	6.14	7.07	10.92
-21.74	697.9											1424.6	2141.2	2871	3586.3	4317.3	5031.9
Na	Mg											Al	Si	P	S	Cl	Ar
2.80	2.6											3.22	4.68	5.62	6.24	8.32	7.7
51.29	1800.7											3540.3	5274.6	7049.7	8880.8	10566.3	12298.7
2.89	4.99											2.81	3.43	4.89	4.16	4.68	8.11
-46.08	1480											2997.4	4510.1	6057.7	7653.9	9124.5	10635.5
K	Ca	Sc	Ti	V	Cr	Mn	Fe	Co	Ni	Cu	Zn	Ga	Ge	As	Se	Br	Kr
2.39	2.29	3.43	3.64	3.85	3.74	3.85	4.26	4.37	4.37	4.47	4.26	3.22	4.58	5.3	5.93	7.59	6.86
81.81	2864	2865.8	2867.5	2869.1	2870.5	2871.8	2872.9	2874	2875	2850.8	2852.5	5650.4	8410.2	11226.2	13978.1	16805.2	19551.4
1.98	3.85	3.22	3.22	2.91	3.12	3.64	3.64	3.43	3.22	3.22	5.2	2.81	3.33	4.47	3.85	4.26	7.28
-73.49	2353.9	2355.5	2356.9	2358.3	2359.3	2360.3	2361.3	2362.2	2363	2343.1	4783.9	7191.2	9646.6	12046.9	14512	16907.3	
Rb	Sr	Y	Zr	Nb	Mo	Tc	Ru	Rh	Pd	Ag	Cd	In	Sn	Sb	Te	I	Xe
2.29	1.98	3.43	3.85	4.06	4.06	3.62	4.47	4.47	4.16	3.72	4.26	4.89	5.51	6.76	7.59	5.82	
107.4	3802.6	3789.3	3777.3	3766.5	3800.3	3789.3	3779.2	3769.9	3798.3	3780.6	7471.1	11172.7	14787.9	18490	22198.5	25911.7	
1.87	3.74	2.91	3.02	2.91	3.12	3.64	3.43	3.22	3.64	3.12	4.78	2.70	3.02	3.85	3.54	3.74	6.34
-96.52	3125.4	3114.5	3104.6	3095.7	3123.5	3114.5	3106.2	3098.6	3122.2	3114.5	3107.3	6325.4	9553.3	12707.1	15935.9	19169.4	22407.5

54   Mihai V. Putz

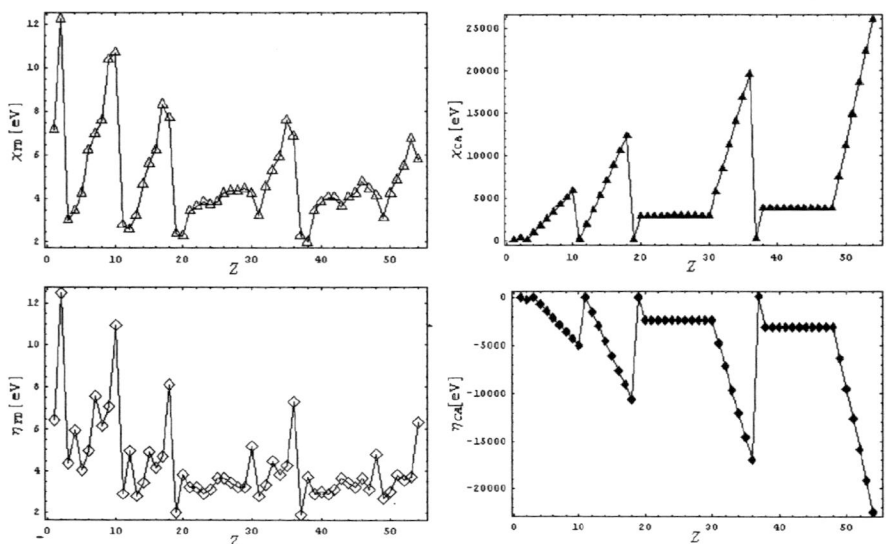

Figure 4. From up to down: the finite difference chemical electronegativity and hardness (on the left column), together with the corresponded chemical action -electronegativity and -hardness (on right column), from table 9, respectively, for the ordinary elements.

Table 10. The chemical action together with limited-, first bosonic and second bosonic electronegativity, calculated with the help of relations (113), (115), and (116) upon equations (85), (97) and (105), respectively, in eV (electron-volts), for the ordinary elements.

				H	He	Legend: Symbol of Element										
				27.21	54.42	Chemical Action: $-C_A$										
				7.18	20.4	Limited Electronegativity: χ_L										
				7.18	2.61	First Bosonic Electronegativity: $\chi_B^{(1)}$										
				7.18	2.1	Second Bosonic Electronegativity: $\chi_B^{(2)}$										
Li	Be										B	C	N	O	F	Ne
91.71	182.48										275.12	365.7	458.53	549.03	641.94	732.4
4.37	9.87										17.40	27.18	38.86	52.74	68.35	86.07
7.61	3.58										2.14	1.32	0.75	0.28	-0.11	-0.48
39.84	11.41										6.44	4.84	4.21	3.88	3.72	3.61
Na	Mg										Al	Si	P	S	Cl	Ar
194.38	387.										578.84	770.29	967.5	1171.75	1356.72	1548.
6.47	10.89										16.46	23.16	30.08	39.3	49.52	60.62
7.61	3.67										2.31	1.59	1.13	0.8	0.53	0.3
61.24	23.12										14.16	10.86	9.41	8.71	8.17	7.87
K	Ca	Sc	Ti	V	Cr	Mn	Fe	Co	Ni	Cu	Zn	Ga	Ge	As	Se	Br	Kr
310.04	615.51	615.91	616.27	616.6	616.9	617.18	617.43	617.67	617.89	612.67	613.04	923.86	1228.2	1540.68	1844.29	2157.79	2460.88
4.28	7.23	8.01	8.82	9.68	10.57	11.5	12.48	13.49	14.54	15.77	16.90	22.21	28.43	35.2	42.91	51.13	60.32
7.71	3.78	3.77	3.75	3.74	3.72	3.71	3.69	3.67	3.66	3.63	3.61	2.29	1.6	1.17	0.87	0.64	0.45
144.55	50.46	46.71	43.47	40.66	38.21	36.05	34.14	32.45	30.94	29.17	27.99	19.88	16.31	14.61	13.56	12.98	12.55
Rb	Sr	Y	Zr	Nb	Mo	Tc	Ru	Rh	Pd	Ag	Cd	In	Sn	Sb	Te	I	Xe
407.18	817.24	814.37	811.79	809.46	816.74	814.37	812.2	810.21	816.38	814.37	812.5	1221.55	1631.63	2029.48	2439.6	2850.29	3261.43
3.38	5.65	6.28	6.94	7.64	8.28	9.04	9.83	10.66	11.41	12.3	13.22	17.42	22.2	27.72	33.67	40.18	47.27
7.74	3.82	3.81	3.80	3.79	3.78	3.77	3.76	3.75	3.74	3.72	3.71	2.39	1.72	1.3	1.01	0.79	0.62
239.84	82.57	75.52	69.5	64.34	61.09	57.06	53.52	50.40	48.45	45.92	43.66	29.58	23.65	20.47	18.79	17.76	17.09

Atomic Electronegativity and Hardness 55

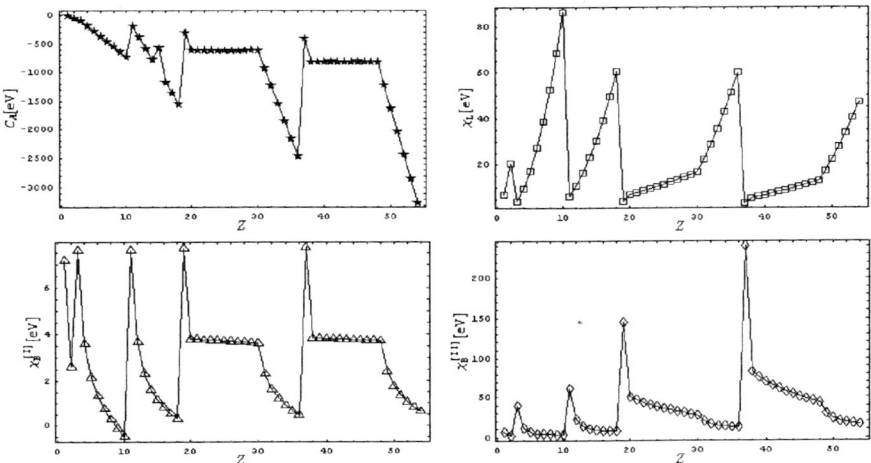

Figure 5. From up to down and left to right: the chemical action together with limited-, first bosonic- and second bosonic- electronegativity, from table 10, respectively, for the ordinary elements.

Table 11. The limited- and limited chemical action- hardness, first bosonic- and first bosonic chemical action- hardness, together with second bosonic- and second bosonic chemical action- hardness, calculated with the help of relations (113), (115) and (116), upon equations (86) and (87), (98) and (99), and (106) and (107), respectively, in eV (electron-volts), for the ordinary elements.

H	He	Legend: Symbol of Element
6.45	26.94	Limited Hardness: η
6.45	12.9	Limited Chemical Action Hardness: η_L^{CA}
6.45	1.61	First Bosonic Hardness: $\eta_B^{[1]}$
6.45	13.22	First Bosonic Chemical Action Hardness: $\eta_B^{[1],CA}$
6.45	0.76	Second Bosonic Hardness: $\eta_B^{[11]}$
-6.45	1.4	Second Bosonic Chemical Action Hardness: $-\eta_B^{[11],CA}$

Li	Be										B	C	N	O	F	Ne
0.66	1.69										3.5	6.46	10.62	16.5	23.98	33.78
21.74	43.26										65.22	86.69	108.7	130.15	152.18	173.62
6.45	1.61										0.72	0.40	0.26	0.18	0.13	0.1
22.17	44.42										67.02	89.11	111.73	133.79	156.43	178.48
36.96	4.71										1.53	0.76	0.49	0.35	0.28	0.23
-39.67	2.17										9.82	14.55	18.8	22.78	26.79	30.67
Na	Mg										Al	Si	P	S	Cl	Ar
0.65	0.93										1.43	2.13	3	4.05	5.57	7.34
46.08	91.74										137.25	182.61	229.36	277.78	321.63	366.97
6.45	1.61										0.72	0.40	0.26	0.18	0.13	0.1
47.19	94.26										141.04	187.7	235.76	285.54	330.62	377.23
55.85	9.47										3.42	1.77	1.13	0.82	0.64	0.52
-57.59	5.2										20.43	30.46	39.54	48.54	56.57	64.79
K	Ca	Sc	Ti	V	Cr	Mn	Fe	Co	Ni	Cu	Zn	Ga	Ge	As	Se	Br	Kr
0.17	0.25	0.31	0.38	0.45	0.54	0.64	0.75	0.88	1.02	1.21	1.39	1.59	1.96	2.40	2.99	3.63	4.44
73.5	145.91	146.01	146.1	146.17	146.31	146.37	146.43	146.48	145.33	145.24	149.26	219.01	291.16	365.24	437.21	511.58	583.38
6.45	1.61	1.61	1.61	1.61	1.61	1.61	1.61	1.61	1.61	1.61	1.61	0.72	0.4	0.26	0.18	0.13	0.1
75.37	149.95	150.05	150.14	150.21	150.29	150.36	150.42	150.47	150.53	149.26	150.65	225.12	299.29	375.44	449.43	525.83	599.69
134.54	21.69	19.87	18.30	16.94	15.75	14.7	13.78	12.95	12.22	11.39	12.47	2.58	1.73	1.27	1.01	0.83	
-145.42	0.45	2.61	4.47	6.09	7.51	8.76	9.86	10.83	11.71	12.47	13.16	33.06	48.86	63.06	76.42	89.98	102.99
Rb	Sr	Y	Zr	Nb	Mo	Tc	Ru	Rh	Pd	Ag	Cd	In	Sn	Sb	Te	I	Xe
0.08	0.11	0.14	0.17	0.21	0.25	0.29	0.35	0.41	0.47	0.54	0.63	0.73	0.89	1.11	1.37	1.67	2.02
96.53	193.74	193.06	192.45	191.89	193.06	192.54	192.07	193.53	193.06	192.61	289.58	386.8	481.11	578.34	675.7	773.16	
6.45	1.61	1.61	1.61	1.61	1.61	1.61	1.61	1.61	1.61	1.61	0.72	0.4	0.26	0.18	0.13	0.1	
99.05	199.11	198.41	197.78	197.21	198.99	198.41	197.4	197.88	197.4	198.41	197.95	297.66	397.6	494.56	594.5	694.6	794.78
225.17	36.35	32.94	30.04	27.55	25.95	24	22.29	20.79	19.82	18.6	17.51	7.12	3.9	2.51	1.82	1.42	1.16
-248.1	-8.3	-4.41	-1.11	1.74	3.94	6.13	8.05	9.73	11.14	12.49	13.69	43.22	64.34	82.78	100.93	118.76	136.43

Overall, since the aufbau principle calls for theoretical reasons in describing the periodicity of the elements theoretical approach and functions have to be involved.

This way, in tables 8-11 are reported the various atomic scales for the present considerate density functionals of chemical action, electronegativity and hardness, see the sets of equations (59)-(60), (85)-(87), (97)-(99), and (105)-(107), respectively, calibrated according with table 7 and computed using relations (113), (115) and (116) with the inputs from table 6; they are compared, when possible, with some experimental counterparts.

For the shake of clearness the same scales are in figures 3-6 represented. We will briefly comment on each of them emphasizing on distinct aspects in foregoing paragraphs.

In table 8 and figure 3 there are jointly presented and compared the first ionization potentials and electron affinities [162] respecting the corresponded HOMO- and LUMO- chemical actions calculated upon equations (57) and (58). One has noted, however, that in calibrating the HOMO chemical action C_A^- and LUMO chemical action C_A^+ the initial values for obtaining experimental EA for H and IP for He have been settled yielding the multiplication factors about $-27.21 \times 7.31 \times 10^{-4}$ and $-27.21 \times 4.94 \times 10^{-2}$ electron-volts, respectively.

The obtained values and scales clearly emphasize on the regular periodic trends of chemical action related functionals respecting their experimental counterparts. Anyway, as previous discussed, here, the chemical action functionals refer exclusively to the intrinsic theoretical characteristics of atoms; the fact that they reproduce the systematic arrangement of the table of elements give enough support for validation them, as a measure of the aufbau principle.

Nevertheless, another important feature regards their apparent inverse albeit periodical trends of C_A^- and C_A^+ compared with those suggested from the experimental IP and EA, respectively. This can be explained by the quantum fluctuation role that chemical action assumes in general, via its basic variational principle – see equation (38), when the kinetic and inter-electronic interactions are neglected. In other words, since the explicit free and exchange-correlation energetic terms are ignored, chemical action takes the role of averaging all these through optimizing the virtual chemical reactivity paths – see equation (46). This way, as the number of occupied shells increases the energy required for stabilizing the electronic fluctuations rises as well, from where the constant up-shifting of the valence response of main group elements.

The same behavior is recorded for the transitional metals, although on more compact range of values due to their close lying chemical nature. In the same context, there are observed the impressive high values for both HOMO and LUMO chemical actions with a more obvious effect on first; this is a natural consequence of effective summation of all associated orbital chemical actions situated under that currently investigated, according with the equivalencies (61) and (62), respectively.

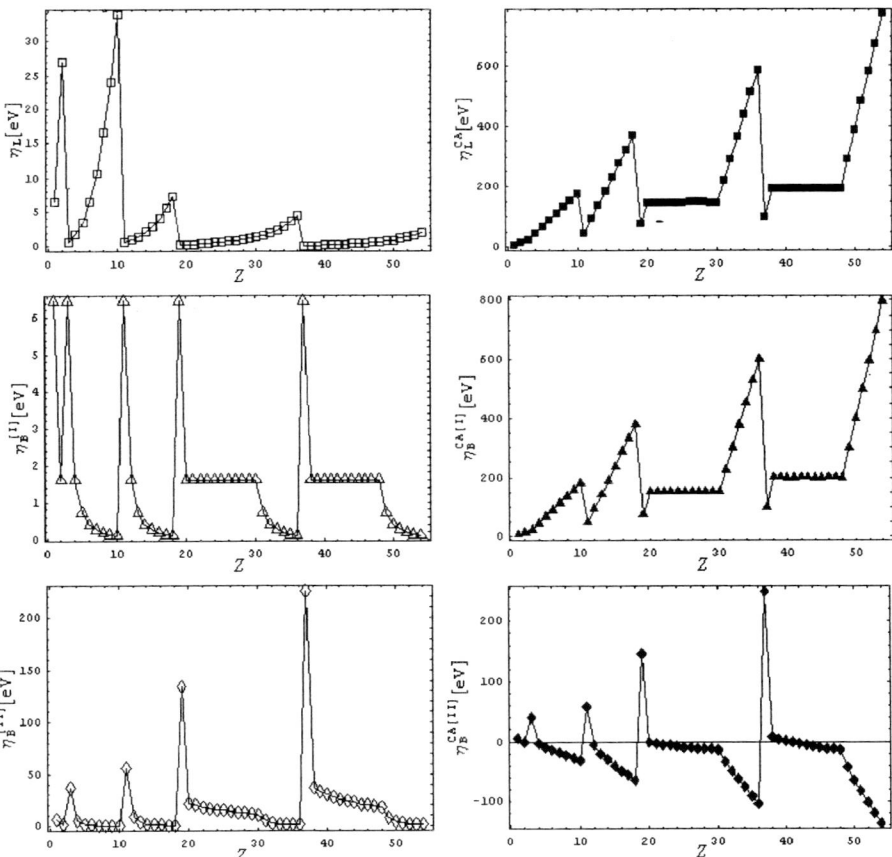

Figure 6. From up to down and left to right: limited- and limited chemical action- hardness, first bosonic- and first bosonic chemical action- hardness, and second bosonic- and second bosonic chemical action- hardness, from table 11, respectively, for the ordinary elements.

Of course, these quantities can not be directly measured but serve for theoretical proof of the elemental periodicity in a clear manner; additionally they can be employed when other intrinsic pure quantum effects as superconductivity or Bose-Einstein condensate are investigated at the atomic level [177, 178].

Instead, in table 9 and figure 4 there are side by side plotted chemical electronegativity and hardness computed on the couples of values (*IP*, *EA*) and (C_A^-, C_A^+) of table 8, upon equations (11) and (13), and (59) and (60), respectively. As before there is much evidence that while finite difference values based on *IP* and *EA* gives many irregularities the chemical action based ones display a smooth periodic fashion. Also the other specificities from HOMO and LUMO chemical action functionals are conserved here: the same constant increasing trend down groups and across periods of main elements instead of the expected decrease between periods; also high ranges of values are recorded.

We have in fact developed new atomic scales in its true sense that a new periodic trend was gained by means of employing quantum effects trough chemical action functionals and principle. As intrinsic values they do not mimic the celebrated values and scales of Pauling, Mulliken or others [19, 51, 83, 112, 179-184], instead acquiring the freedom to relate merely with the quantum structure rather than with the experimental quantities.

Nevertheless, here, the chemical action hardness poses, apart of the new periodic trend, also the almost overall negative sign excepting the alkaline elements that represents the turning point to a new period. This can suggests that in these conditions the chemical hardness provides an extra-quantum effect when actively stabilizes the atoms in the molecule. Putted differently, the chemical action hardness has "the power" not only to dissipate the remaining unsaturated hardness through the HSAB equalization but also to dissolve somehow the atoms in the formed molecule. This case can cover the situations when atomic partitions of molecules are meaningless for interpreting reactivity; that means that the quantum fluctuations dominates the individuality of atoms in molecule thus destroying their basins or lodges [185, 186].

With the table 10 and figure 5 we turn to the analysis of the density functional softness based electronegativity, in its limited-, first bosonic and second bosonic realizations, see relations (85), (97), and (105), respectively. However, EN functional values are compared with simple chemical action (35) ones. Now, as observed, the chemical action displays the negative range of values due to its calibration factor $27.21 \times 5.45 \times 10^{-2}$ eV that fixes C_A to double ground energy of H atom, according with relation (48) and the virial relation (49), when neglecting the inter-electronic interaction. All previous HOMO and LUMO chemical action characteristics are present also in the chemical action scale.

However, the three computed versions of electronegativity display different scales to each other. The limited local softness based one χ_L behaves almost like finite-difference experimental based one χ_{FD} of table 9 showing that apart displaying more regulate trend does not account on particular quantum fluctuation effects. Contrarily, the first and second bosonic electronegativities brake with the trend of the limited- and chemical action- based ones by presenting one more way of periodicity of elements: increasing down groups still decreasing through periods.

These new types of atomic scales are consistent with the fermionic-bosonic mixture conditions in which their (97) and (105) density functionals have been derived. Since the electronegativity scale without quantum fluctuation effects displays a normal trend, i.e. the decrease down groups and increase along periods, the present situation is an up-side-down one. Again, this is due to the presents of the indirect chemical action influence, see relation (88), which is nevertheless partly removed when taking the asymptotic limits (95) and (96). With these, the present bosonic scales can be at best used when assaying the atomic systems to from molecular samples and bonds in conditions that circumvent the Bose-Einstein quantum condensation effects.

Finally, in table 11 and figure 6 there are collected and represented the values of the limited-, first bosonic and second bosonic- hardness and of the associate scales when chemical action contribution was added, based on the equations (86), (98), (106) and of (87), (99), and (107), respectively. As in the case of electronegativity formulations based on density functional softness theory, the limited hardness η_L reflects the periodic scale without dominant quantum fluctuation effects thus recovering the basic finite-difference trend of experimental hardness η_{FD} of table 9 and figure 4. As well, the bosonic scales of hardness are in parallel with those of electronegativities of table 10 and figure 5.

As already anticipated, when the chemical action is directly involved it dominates the scales thus producing the counterpart limited-, first bosonic and second bosonic chemical action hardness scales, displayed on the right column of figure 6, with basically the same characteristics as those showed by the chemical action related functionals on the right columns of figure 3 and 4. Notably, the second bosonic chemical action hardness $\eta_B^{CA[II]}$ plays here the role of the simple chemical action hardness η_{CA} of figure 4 in the sense of posing almost all negative values, with the effect in quantum destroying of the atomic identity in potentially formed molecules. Nevertheless, in the context of Bose-Einstein condensation this atomic feature becomes a crucial one for the new created many-electronic poly-atomic state.

Still, although with different meanings and significance, the above electronegativity and hardness scales worth to be engaged in questing for the conditions in which the quantum nature of the chemical bond is manifested via the electronegativity and hardness principles of reactivity, see table 1, and at which degree. This attempt will be in what next illustrated.

Chapter 6

6. MOLECULAR ELECTRONEGATIVITY AND HARDNESS

Since molecules are made from atoms that preserve most of their identities apart of the valence or frontier interactions the natural way of treating the bonding indices is to compose them in an iterative manner from those that virtually characterize the isolated atoms. Therefore, the molecular electronegativity and hardness can be built upon atomic electronegativity and hardness by appropriate rules. That rules have to closely follow the bonding principles, more exactly the electronegativity equalization principle and HSAB principle as well. These principles are the first called since, in fact, both provide the equalization of electronegativity and hardness in bonds, see table 1, while their inequality variants are merely associated with the quantum fluctuation effects that still remain after the equalization principles are consumed. As such, the main working principles are electronegativity and hardness equalization ones to provide the molecular or bonding electronegativity and hardness. The general procedures to derive them as well as illustrative application to chemical reactions are in what follow presented.

6.1. GENERAL ALGORITHM

The starting point is to treat the diatomic molecule case AB. In this respect, let's consider that the formation of the diatomic molecule AB takes place with constant atomic nuclear charges at the equilibrium separating distance R_{AB}. To globally design the binding the infinitesimal electronic charge transfers between the molecule's atoms is considered to adjust the atoms-in-molecule resulting charges, respectively as:

$$N_{\langle A \rangle} = N_A + dN_{\langle A \rangle},$$
$$N_{\langle B \rangle} = N_B - dN_{\langle B \rangle} \qquad (117)$$

Consequently, at the energetic level, the variation or the total energy of the $A+B$ system can be written as:

$$dE = dE_{\langle A \rangle} + dE_{\langle B \rangle}$$

$$= \left(\frac{\partial E}{\partial N_{\langle A \rangle}}\right)_{N_{\langle B \rangle}, R_{AB}} (N_{\langle A \rangle} - N_A) - \left(\frac{\partial E}{\partial N_{\langle B \rangle}}\right)_{N_{\langle A \rangle}, R_{AB}} (N_{\langle B \rangle} - N_B) + \left(\frac{\partial E}{\partial R_{AB}}\right)_{N_{\langle A \rangle}, N_{\langle B \rangle}} dR_{AB} \qquad (118)$$

Now, at equilibrium the bonding state is fixed by the differential conditions:

$$\frac{\partial E}{\partial R_{AB}} = 0, \qquad (119)$$

$$dE = 0, \qquad (120)$$

thus leaving from (118) with the electronegativity equalization phenomenology,

$$\left(\frac{\partial E_{\langle A \rangle}}{\partial N_{\langle A \rangle}}\right)_{N_{\langle B \rangle}, R_{AB}} = \left(\frac{\partial E_{\langle B \rangle}}{\partial N_{\langle B \rangle}}\right)_{N_{\langle A \rangle}, R_{AB}}, \qquad (121)$$

through recognizing the electronegativity basic energetic definition (10).

Next, by employing this principle on the atoms A and B in AB, i.e. by applying the equation (121) on the parabolically exchanged charge truncated energetic expansion, see equation (19), one yields that the electronegativities of atoms in molecule, $\chi_{\langle A \rangle}$ and $\chi_{\langle B \rangle}$, equalize to the bonding electronegativity χ_{AB},

$$\chi_{\langle A\rangle} = \chi_A - 2\eta_A \Delta N = \chi_{\langle B\rangle} = \chi_B + 2\eta_B \Delta N = \chi_{AB}, \qquad (122)$$

from which equivalent expressions the number of the transferred charges immediately casts as:

$$\Delta N = \frac{\chi_A - \chi_B}{2(\eta_A + \eta_B)} \qquad (123)$$

However, before going further worth noting that from relation (123) appears that the electronegativity difference is crucial for binding promotion and bonding formation, confirming once again that the competition between electronegativities of atoms in binding a molecule is the true meaning of the famous Pauling statement that the electronegativity of an atoms is "the power with which an atom in a molecule attract the electrons to itself" [19]. The Pauling definition does no regard isolated atom electronegativity but the equalization electronegativity principle in molecular samples – see also equation (2) and the accompanied discussion.

With (123) back in whatever part of (122) the average value of the equalized electronegativity of atoms in diatomic molecule AB is released:

$$\chi_{AB} \equiv \overline{\chi} = \frac{\eta_A \chi_B + \eta_B \chi_A}{\eta_A + \eta_B} \qquad (124)$$

Still, aiming to obtain from (124) also the general formula for n^{AIM} atoms in a molecule say, we have to rewrite it as much as possible without the direct chemical hardness involvement. This goal can be achieved if one observes that electronegativity and hardness go somehow "hand in hand" in establishing the bonding scenario of table 1, as well as there is suggested by their associate atomic scales of figures 4-6 (with the best emphasis on the figure 4 where the chemical action related electronegativity and hardness scales are represented). Therefore, a sort of universal invariant θ of their proportion can be assumed such that for every atoms and molecule the relations hold:

$$\chi_A = \theta \, \eta_A, \chi_B = \theta \eta_B, \qquad (125)$$

$$\chi_{AB} = \theta \eta_{AB} \qquad (126)$$

Replacing the rules (125) in (124) the simple diatomic averaged electronegativity is primarily obtained,

$$\overline{\chi} = \frac{\dfrac{\chi_A}{\eta_A} + \dfrac{\chi_B}{\eta_B}}{\dfrac{1}{\eta_A} + \dfrac{1}{\eta_B}} \stackrel{(125)}{=} \frac{2}{\dfrac{1}{\chi_A} + \dfrac{1}{\chi_B}}, \qquad (127)$$

that from its mathematical structure permits from now the direct generalization to the polyatomic molecules throughout the formula [28, 29]:

$$\chi_{M \pm \delta} = \frac{n^{AIM} \pm \delta}{\displaystyle\sum_A \frac{n^A}{\chi_A}} \qquad (128)$$

In expression (128) the sum of the n^A atoms of each species present in the molecule recovers the total atoms in molecule n^{AIM},

$$\sum_A n^A = n^{AIM}, \qquad (129)$$

being also the overall molecular charge $\pm \delta$ included in the formula, for the shake of completeness.

Turning to the associate molecular hardness formulation, the relations (125) are now employed to replace the individual electronegativities from the basic formula (124) that by further combination with the molecular invariant relation (126) results in the diatomic averaged hardness:

$$\eta_{AB} \equiv \overline{\eta} = 2\frac{\eta_A \eta_B}{\eta_A + \eta_B} = \frac{2}{\dfrac{1}{\eta_A} + \dfrac{1}{\eta_B}} \qquad (130)$$

By generalizing the expression (130), it leads with the polyatomic hardness in the same spirit as electronegativity (128) was previously formulated:

$$\eta_{M^{\pm\delta}} = \frac{n^{AIM} \pm \delta}{\sum_{A} \dfrac{n^{A}}{\eta_{A}}} \qquad (131)$$

with relations (128) and (129) the different atomic scales for electronegativity and hardness can be used to generate their molecular averaged counterparts in assisting the interpretation of the chemical reactions. Concrete examples are in what next analyzed with the help of the actual unified atomic chemical action-, limited-, first- and second- bosonic related electronegativity and hardness.

6.2. APPLICATION TO CHEMICAL REACTIONS

One of the very purposes of the modern conceptual chemistry stands the capacity of modelling and controlling of the chemical reaction via theoretical methods.

There is also recognized that only with admission of the electronegativity and hardness concepts the chemical reactivity, and the chemistry in general, has the benefit to pose its own general principles, see table 1, based merely on mathematical rather on physical laws. There is therefore desirable to apply the electronegativity and hardness principles and of their consequences in a unitary manner at whatever level of computational, in principle, for any kind of chemical reaction.

At this point, the polemic especially around the chemical hardness principles has been arisen [142, 151]. While the mathematical proof of MHP was, phenomenologically, grounded on or related with the second principle of thermodynamics and links with the growing of entropy in open chemical systems [4, 117], the HSAB principle was often confused with the Pearson classification of Lewis acids and bases. Of course, the HSAB implies the Pearson classification, in some circumstances, although is not derived from that.

Recalling the previous identified hardness nature as the chemical force, see equation (12) and the subsequent discussion, it means that the chemical reactivity implies that a *chemical equilibrium* is reached within each formed molecule emerged out of a reaction. More, this chemical equilibrium consists in two steps:

one regarding the chemical potential equalization (electronegativity equalization principle) followed by the chemical force equalization condition (hard with hard and soft with soft acids and bases principle).

The fact that in many situations the computations do not recommend the hard and soft classification of acids and bases as prescribed by the Pearson classification, or by some experience, has to be granted to the quantum fluctuation effects, again both at electronegativity (as originator of hardness) and hardness levels, through the chemical potential inequality and maximum hardness principles, see table 1.

Therefore, we advocate for assuming as foreground chemical reactivity principles those calling on electronegativity and hardness, on whose basis specific analysis to be performed and interpreted accounting on the degree with which the quantum effects were included.

In next, a recent well studied chemical reaction will be taken as the working example due to its enough general character but also for the problematical issues reported with occasions of its previous investigation [74]. The general scheme is based on the reaction between a substrate $(X-H^+)$ containing a replaceable group X (a base) bounded to the hard electrophilic Lewis acid (H^+) and the HO–OH acid-base complex of the hard base (OH^-) and a soft acid (HO^+) [74]:

$$H - X^+ + HO - OH \leftrightarrow HO - X^+ + H - OH \qquad (132)$$

The problem is to establish the belonging class of the group X when it varies through a series of bases in chemical reaction (132). The engaged testing bases are initially grouped according with the Pearson classification as hard and soft, as presented in table 12.

The previous criterion for hard and soft classification was suggested by Pearson to be the reaction energy ΔG [4, 68]: the more negative the reaction energy the softer base would be, a rule also used in table 12 for classification of the bases under study [74].

Table 12. Qualitative classification of Lewis bases tested in this work [4, 74].

Soft	Hard
$CN^- < CH_3S^- < CH_3SH$ $<$	$CH_3O^- < NH_3 < H_2O$

Molecular Electronegativity and Hardness

Table 13. Unified absolute and chemical hardness, computed at the experimental (Exp), chemical action (CA), limited (L), limited chemical action ($_L{}^{CA}$), first bosonic ($_B{}^{[I]}$), first bosonic chemical action ($_B{}^{CA[I]}$), second bosonic ($_B{}^{[II]}$), and second bosonic chemical action ($_B{}^{CA[II]}$) levels of approximations, by employing the tables 9-11 to the general molecular hardness formula (131), together with the Hartree-Fock (HF) estimated chemical hardness and reaction energy of (132), for the species of table 12.

Species	ΔG_{HF}^*	η_{HF}^*	η_{Exp}	η_{CA}	η_L	η_L^{CA}	$\eta_B^{[I]}$	$\eta_B^{CA[I]}$	$\eta_B^{[II]}$	$\eta_B^{CA[II]}$
CN⁻	-216.7	9.01	3.01	-1226.48	4.02	48.23	0.16	49.57	0.3	-8.2
CH₃S⁻	-181.2	6.5	4.42	8.61	4.61	8.33	0.47	8.34	1.33	10.64
CH₃SH	-78.7	7.33	5.66	9.68	5.87	9.44	0.69	9.45	1.90	11.3
CH₃O⁻	-23.4	7.21	4.83	8.61	5.88	8.26	0.47	8.27	0.86	11.35
NH₃	-17.2	8.44	6.70	8.61	7.15	8.43	0.93	8.44	1.6	9.71
H₂O	61.5	9.59	6.34	9.68	8.09	9.44	0.51	9.45	0.95	11.27

*values abstracted from [74].

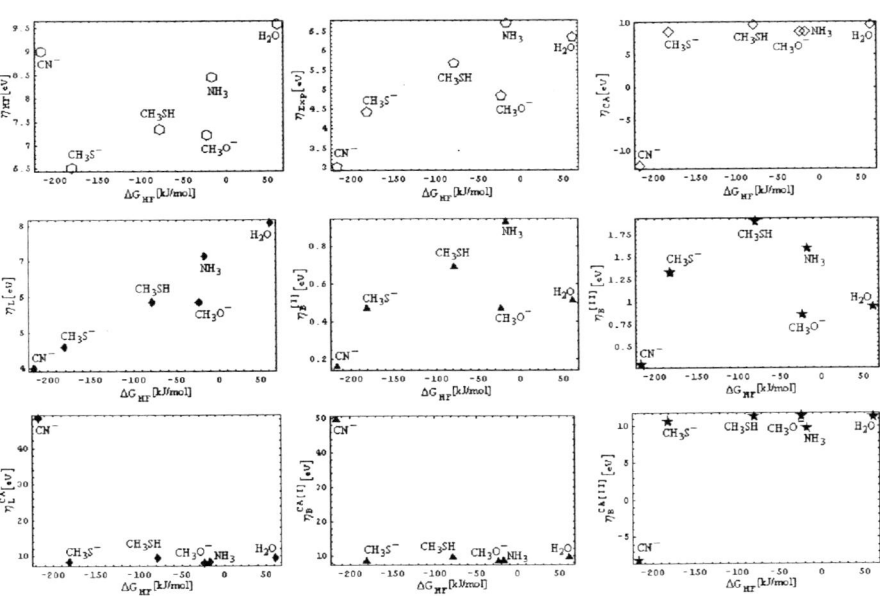

Figure 7. Graphical correlation of the molecular chemical hardness values respecting the negative of the reaction energy of (132), as reported in table 13, for the bases of table 12. The η_{CA} value of CN⁻ was intentionally scaled to −12.26eV to can be adequately represented respecting the other values, although its value of table 13 is the true computed one.

Table 14. Unified absolute and chemical electronegativity, computed at the experimental (Exp), chemical action (CA), limited (L), first bosonic ($_B^{[I]}$), and second bosonic ($_B^{[II]}$), levels of approximations, by employing the tables 9-11 to the general molecular electronegativity formula (128), together with the Hartree-Fock (HF) estimated electronegativity, for the species of table 12.

Species	χ_{HF}^*	χ_{Exp}	χ_{CA}	χ_L	$\chi_B^{[I]}$	$\chi_B^{[II]}$
CN$^-$	-4.86	3.29	1431.36	15.99	0.48	2.25
CH$_3$S$^-$	-4.74	5.42	9.56	8.33	1.65	5.41
CH$_3$SH	2.54	6.84	10.76	9.69	2.34	6.83
CH$_3$O$^-$	-5.38	5.64	9.56	8.45	0.84	4.53
NH$_3$	2.21	7.13	9.57	9.02	2.28	6.1
H$_2$O	3.96	7.31	10.76	10.08	0.78	5.59

*values abstracted from [74].

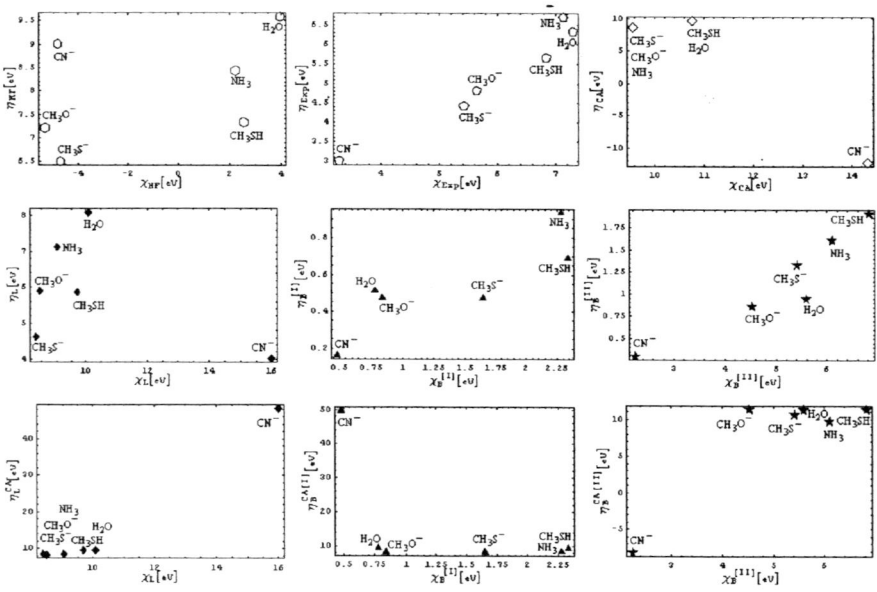

Figure 8. Graphical correlation of the molecular chemical hardness values respecting the molecular electronegativity, as reported in tables 13 and 14, for the bases of table 12. The η_{CA} and χ_{CA} values of CN$^-$ were scaled to -12.26 and $14.31 eV$ to can be adequately represented respecting the rest of values, although their values of table 13 and 14 are the true computed ones, respectively.

Molecular Electronegativity and Hardness

Worth noting that within even the most sophisticated density computational versions for evaluating the energies of reactions for molecules in table 12, combined with the most improved schemes for extending chemical hardness evaluation beyond the second order central difference of (13) up to the spectral like resolution, the output of computed hardness have shown poorly correlation with the prescribed hard-soft order paralleling the increase of the negative of the reaction energy [74].

Table 15. Maximum hardness index (25) computed at the experimental (Exp), chemical action (CA), limited (L), limited chemical action ($_L^{CA}$), first bosonic ($_B^{[I]}$), first bosonic chemical action ($_B^{CA[I]}$), second bosonic ($_B^{[II]}$), and second bosonic chemical action ($_B^{CA[II]}$) levels of approximations, by employing the tables 9-11 to the iterative molecular hardness formula (131), along with the Hartree-Fock (HF) estimated one based on HF chemical hardness values of table 13, for the species of table 12.

Species	Y_{HF}	Y_{Exp}	Y_{CA}	Y_L	Y_L^{CA}	$Y_B^{[I]}$	$Y_B^{CA[I]}$	$Y_B^{[II]}$	$Y_B^{CA[II]}$
CN⁻	0.994	0.945	1.	0.969	0.999	-18.53	0.999	-4.56	0.993
CH₃S⁻	0.988	0.974	0.993	0.976	0.993	-1.26	0.993	0.717	0.996
CH₃SH	0.991	0.984	0.995	0.985	0.994	-0.05	0.994	0.86	0.996
CH₃O⁻	0.99	0.979	0.993	0.986	0.993	-1.26	0.993	0.32	0.996
NH₃	0.993	0.989	0.993	0.99	0.993	0.42	0.993	0.80	0.995
H₂O	0.995	0.988	0.995	0.992	0.994	-0.92	0.994	0.45	0.996

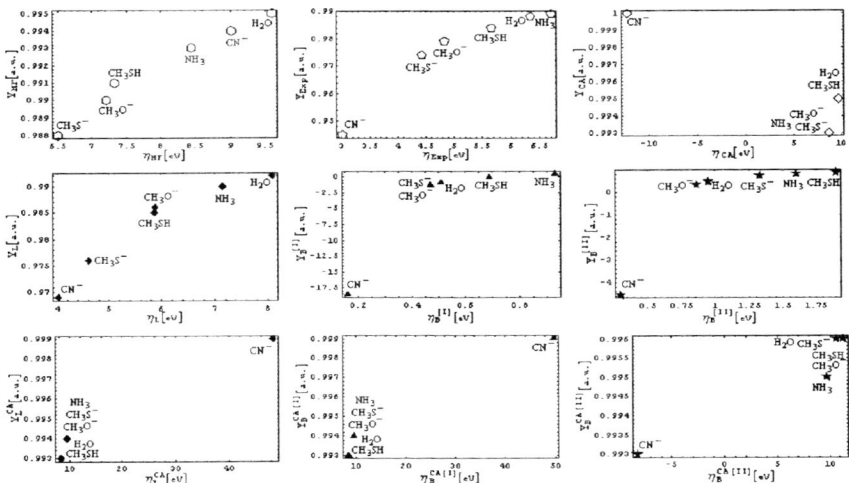

Figure 9. Graphical correlation of the maximum hardness indices respecting the chemical hardness values, as reported in tables 13 and 15, for the bases of table 12.

However, we claim that this does not mean that the HSAB principle is not true, being the controversial results an indication that the counterpart quantity with which the hardness to correlate has to be changed, or the quantum fluctuation effects to be included. In what follows we will search, nevertheless, through the unified forms of hardness and electronegativity of the present approach for those that produce best correlation with the hardness of the molecules in table 12.

The equalization principles of electronegativity and hardness are included since the molecular electronegativity and hardness are computed upon the relations (128) and (131) while the quantum effects and, consequently, their corresponding inequality principles of table 1 are compressed in the different levels of chemical action involvement through their employed density functionals, see tables 3-5 and the adopted unified form from them.

With these remarks the tables 13-15 and the associate figures 7-9 are produced whose meanings and interpretations regarding hard and soft ordering of the focused chemical species are separately discussed.

Figure 7 represents the correlation of the hardness values of table 13 with the associate energies of reaction (132) for the Lewis bases of table 12.

The Hartree-Fock graph reproduce the earlier results [74], showing the complete disconnection of the hard and soft classification respecting the systematic growth of energy of type reaction (132). This because there is clear that the Hartree-Fock scheme, but also all other computational combinations [74], leaves with the conclusion that the softest and hardest X groups in (132), CN^- and H_2O, respectively, display quite similar hardness positions although at extremes of the computed ΔG.

Such behavior is no more met when the present unified scheme of hardness computations are adopted. Maybe not surprising, the best hierarchy is obtained when chemical action hardness atomic scale is used in computing the molecular hardness of species in table 12.

The chemical action related hierarchy, firstly, correctly situates the extreme hard and soft species on the opposite points of depicted graphs; secondly, it regulates the pictures on which is added with the specific sigmoid shape of the chain of ordering molecules.

However, an output in agreement with table 12 is obtained only with η_{CA} and $\eta_B^{CA[II]}$ scales for the concerned molecular species, whereas the scales η_L^{CA} and $\eta_B^{CA[I]}$ practically reverse the recommended classification. These last trends can be rationalized remembering that the limited and first bosonic cases are based on *ab initio* truncated working density functional expressions on which when further fluctuations are considered may provided complete inverse reactive trend. The last assertion is sustained also from their atomic scales of figure 6, a situation somehow remedied for the second bosonic hardness $\eta_B^{CA[II]}$.

Nevertheless, as from the figure 6 was also predicted, the atomic η_L scale, in accordance with the common rules of electronegativity related scales, i.e. decreasing down groups-increasing through periods [13], provide an almost correlated scheme between the hardness of the species of table 12 and the associated ΔG reaction energies of table 13. It shows an even better correlation than that depicted on the experimental based hardness values η_{Exp}.

Therefore, there was in any case proved that proper hardness models furnishing an acceptable correlation with the reaction energies do exist and they depend on the specific quantum constraints adopted.

Going to the figure 8 based on the table 14, we are in front of a new type of correlation, a hardness-electronegativity one, at the molecular level. This kind of analysis seems also natural since the basic chemical reactivity principles are coined on electronegativity and hardness, as the bonding scenario from table 1 clearly suggests. Still we have to interpret the correlation of figure 8 in the light of right-hand side diagram of figure 2.

Applying the conceptual matrix of the right-hand side of figure 2 on the diagrams of figure 8 one can interpret that the up-left...down-right and down-left...up-right diagonals cover the hard-soft...soft-hard and soft-soft...hard-hard molecular combinations of the general acid-base reaction (27), respectively. Consequently, we will say that the first of above correspondences are on the diagonal of reactants while the second set of correspondences associates with the diagonal of products. This way we succeed to rationalize the reactivity within the chemical space (χ, η).

In this context we further see, again, that the Hartree-Fock hardness values provide diagrammatic positions that belong to both above reactive diagonals so offering no clear rationalization of the preference order in bonding. From all other graphs just $\eta_B^{CA[I]}$ and η_L representations display somehow the two diagonals albeit that of the products (soft-soft...hard-hard) is not entirely appeared. We have to understand that in the virtue of the equilibrium between the two sides of general reaction (27) the conceptual diagrams of reactants and products mean that the equilibrium is shifted to one or to other side toward molecular stabilization. With this diagrammatic conceptualization of the chemical reaction respecting the diagonals of the (χ, η) chemical space, one observes that only the scale η_{CA}

arranges the bases of table 12 on the diagonal of reactants thus predicting that the reaction (132) will not proceed to its right end due to the quantum electronegativity and hardness fluctuations.

Instead, when the scales $\eta_{Exp}, \eta_L^{CA}, \eta_B^{[I]}, \eta_B^{[II]}$, and $\eta_B^{CA[II]}$ are adopted the concerned bases are found on the diagonal of products thus allowing the reaction (132) to end at its right side. Nevertheless, from these scales, η_L^{CA} and $\eta_B^{CA[II]}$ show inverse trend from the reasons explained before, when discussed their atomic periodicities, while $\eta_B^{[I]}$ and $\eta_B^{[II]}$ provide a similar dispersion unless the H_2O and NH_3 species take different positions in (χ, η) chemical space. Still, in these conditions, regarding the order of table 12 the representations associated with η_{Exp} and $\eta_B^{CA[II]}$ appear to correlate at best the electronegativity with hardness and with hard and soft classification albeit the places of CH_3SH and CH_3O^- are reversed.

Nevertheless, the real gain of the chemical space approach consists in analyzing the reactivity in direct correlation with the indices that drive the bonding processes, being this way a consistent structural analysis while the dependency on the specific conditions of reactions is assured through assuming different density functionals schemes for them.

Finally, worth introducing an even more way of reactivity interpretation, this time with hardness as a variable of the maximum hardness index (25). The table 15 and figure 9 present the results. The key of analysis is now the maximum hardness index: as its value approaches unity as the hardest values is provided.

Quite surprisingly, event this way of representation does not reproduce the prescribed order of table 12 when the Hartree-Fock or experimental based schemes of computation are adopted, despite the almost collinear correlation from the figure 9. Instead, now there it seems that only the limited scale η_L furnishes the best results, being in fact that one that fully recovers both acceptable correlation and the soft-hard order of table 12. Let's also note, as already mentioned, that the scales η_{CA}, η_L^{CA}, and $\eta_B^{CA[I]}$ reverse hard with soft values in the same quantum spirit in which their atomic scales of figures 5 and 6 behave. For the rest of the scales there are turned out that the maximum hardness index procedure does not allow decisive grouping in the Pearson sense of table 12; however, it opens the way of a new conceptual perception when providing the same values for certain Lewis reagents.

A special remark has to be done on the negative values reported for the maximum hardness values of the first bosonic scale and for the second bosonic value of the CN⁻ molecule in table 15. The significance of such outputs corresponds to the stabilization of the reaction (132) merely to its left-right side as prescribed by the general interpretation of the maximum hardness index definition (26) respecting the paradigmatic acid-base reaction (27). More phenomenological, that means that the first bosonic conditions are not the appropriate one for allowing the acid-base reactions to flow, due to their special fermionic-bosonic limiting state, see the relation (98). Very important, also at this level, the regulating quantum role of the chemical action is reaffirmed as its presents in both first bosonic- and second bosonic- chemical hardness provide values for maximum hardness index that closely approach the unity, see $Y_B^{CA[I]}$ and $Y_B^{CA[II]}$ of table 15.

At the bottom line, much work is still necessary to obtain a complete picture of bonds in isolated and reactive circumstances, through equilibrium and non-equilibrium states, employing a minimal set of indices and principles. The electronegativity and hardness, and their associated principles have furnished enough support to demonstrate that they can be combined in a variety of density functionals forms and ways of correlations to believe that one day they will be able to solve the quantum nature of bonds and bonding.

Chapter 7

7. CONCLUSION

From the whole history there was the privilege of the humankind to invent concepts to modelling and controlling reality and natural phenomena. However, as the Romanian philosopher Constantin Noica (1909-1987) eventually said, one particular history of ideas can be called a discipline if it has classicism, i.e. if it has ancient originators and continuators.

That is the case of quantum theory of matter. There is noted that the father of modern sciences, Aristotle, in his search for the very causes of the Nature writes the *Physica* as the first treaty of the conjugated laws of observed things and processes. However he had suffered for failing in coherently linking his representative true from the first engine of the universe. This way, he was forced to imagine the *Metaphysica* scenario that consecrates the world of virtual causes as the motor of the manifested life.

Many centuries after, Newton gave to the world his *Principia* that further arrange the causes of nature in some simple principles that govern both the earth and out-earth equilibrium states and movements. Nevertheless, at that moment the involved mathematics had called for a proper philosophic or metaphysical foundation. This way, Kant had collected in his *Critics* the main ideas of Newton and distilled them until the mind categories and the rationale of things emerged out from a vast sea of hidden yet actively classes of virtual representability of objects and concepts. It was the time that even the concept of idea was formulated in an abstract manner thus given freedom to ideas to reproduce and combine themselves without afraid that their results depart from some observed or unobserved reality. That was the key from which the modern science was founded on rigorous theoretical bases.

Still, with the dawn of the XX century, or more precisely with the advent of the relativity and quanta concepts in physics, the metaphysics of structure matter had been again manifested through another cornerstone scenario, namely the *Seit und Zeit* opus of Heidegger, from which the time-space-being concepts unfold and fold reality between the quanta and waves of perceptions, measurements and thoughts.

The impact of this history was immense since from the appearance of the quantum theory of matter almost each coherent virtual idea can find its proper place and realization soon or latter. Step by step, even the science branches as chemistry and biology, usually designated to assist the most visible processes and transformations of things, have acquired the quantum nature as the atoms and molecules were recognized to be at the foreground of each life form.

In this respect, Pauling's *Nature of the Chemical Bond* stands as the first collection of convergent-emergent quantum ideas in modern chemistry and biology. It is not a definitive picture of atomic matter, of course, but it has the huge merit to give the way in which the mathematical and physical quantum ideas can be combined to provide the iterative stair of matter organization. It also opens the route to understand the chemical reactivity and biological activity in terms of simple quantum concepts among which the electronegativity had been proved to be of the most help in characterizing chemical reactivity and structure.

There is equally true that the electronegativity, as every insightful concept, was a highly controversial idea of Pauling due to its virtual nature. Still, the electronegativity concept had resisted, and – more – had been always refined through all quantum theories that continuously appeared, culminating with the most simple definition and interpretation as the minus the chemical potential of a system within density functional theory. Consequently, it had permitted the introduction of another quantum quantity, the chemical force or hardness together with which provide the sufficient base for rationalization of the chemical bond.

Because of the complementary quantum nature of electrons, as waves and particles, their coupled behavior can not be seen in an ordinary experimentally way in a chemical bond. Even the atomic radius can not be precisely defined as there is not a definite border of being for the frontier electrons, spanning, in principle, whole space to infinitum. For this reason the treatment of bonding is compulsory to be done at the quantum level in order to understand the inner causes and to anticipate the potential effects in reactivity.

Electronegativity and hardness appear in this frame to generate the minimal and sufficient set of global parameters that assists chemical bonding and reactivity in various chemical and physical conditions. Few of them were in this study presented leaving with unification of electronegativity and hardness absolute and chemical levels of their expressions like density functionals formulations. As such, the chemical action-, local limited- first- and second- bosonic formulas were found with the help of which the molecular aggregates and chemical reactions appears to be solved in many ways, strongly related with the physical conditions in which they have been build.

The chemical action was introduced as the proper link between the orbital and global pictures of electronegativity and hardness due to its proper definition that univocally associates the electronic density with external potential applied on an electronic system, in accordance with the main density functional theorems.

However, electronegativity and hardness stand within the minimum dimensioned set of global indices that characterize bonding and reactivity as the electronic density and the effective applied potential function closely relate with the inner structure of atoms and molecules. This study advocates that a proper combination between these two sets of global and local indices can generate a whole plethora of density functionals with a role in quantifying the many-electronic structures and their transformation at the conceptual rather computational quantum level of comprehension. This was proved though applying the obtained electronegativity and hardness atomic scaled to selected problematical chemical reaction to provide the prediction of reactivity and stabilization of bonds in accordance with the main principles of chemistry: equalization and inequality of electronegativity and hardness, known as the electronegativity equalization, inequality of chemical potential, hard and soft acids and base, and maximum hardness principles, respectively.

In this context, a novel reactivity index for quantifying the maximum of hardness realization was proposed with reliability proved throughout providing the hierarchy for a series of hard and soft Lewis bases. In all these, once again, the chemical action influence was appeared to play the role of averaged quantum fluctuations that stabilize the molecules at the end of bonding process. There was also for the first time indicated the appropriate complete bonding scenario based exclusively on the correlated quantum quantities and principles of the electronegativity and hardness. Therefore, there is still hope that the present scenario will be accompanied by some advanced ultra fast frozen movie of atomic encountering in bonding.

Finally, with these, we can faithfully respond to Kant's famous inquiry dying that the world between "above staring sky and our in-depth moral law" is filled by the nature's pure chemistry: the quantum chemistry.

ACKNOWLEDGMENTS

This work has been motivated during many years by my dear teacher, now my colleague and friend, Prof. Dr. Eng. Adrian Chiriac from Chemistry Department of West University of Timisoara, Romania. It was also possible due the continuous support of my family to which I sacrificed many shining moments. Also, I like to thank Prof. Dr. Dulal Ghosh from the Kalyani University of India for filling me through last years with his wise and springing spirit and also for providing me with some of fundamental references on which this study is based. As well, I whish to address special thank Profs. Dr. Ahmed Zewail, Mircea Diudea, Paul Mezey, Eduardo Castro, Ante Graovac, Lionello Pogliani, Milan Randić, Nenad Trinajstić, Hagen Kleinert, and Axel Pelster, among many other international personalities, for their kind in sharing and exchanging with me the conceptual tools of quantum and chemical bonding throughout fruitful discussions with occasions of various formal and informal meetings. Finally, but not at least, the Romanian National Council of Scientific Research in Universities (CNCSIS), Romanian Chemical Society, European Society of Mathematical Chemistry and American Chemical Society are equally thanked for their continuous support for fundamental exploring studies in Chemistry as a Great Science. The ideas of this work are however dedicated to all those who questing for the true nature of bonding find its quantum character as the viable most precious vehicle.

REFERENCES

[1] Lewis, G. N. *J. Am. Chem. Soc.* 1916, 38, 762-785.
[2] Bamzai, A. S.; Deb, B. M. *Rev. Mod. Phys.* 1981, 53, 95-126.
[3] Sen, K. D.; Jørgensen, C. K.; Eds.; *Electronegativity*; Structure and bonding 66; Springer Verlag: Berlin, 1987.
[4] Sen, K. D.; Ed.; *Chemical Hardness*; Structure and bonding 80; Springer Verlag: Berlin, 1993.
[5] Parr, R. G.; Yang, W. *Density Functional Theory of Atoms and Molecules*; Oxford University Press: New York, 1989.
[6] Kryachko, E. S.; Ludeña, E. V. *Energy Density Functional Theory of Many-Electron Systems*; Kluwer Academic Publishers: Dordrecht, NL, 1990.
[7] March, N. H. *Electron Density Theory of Atoms and Molecules*; Academic Press: London, UK, 1992.
[8] Nalewajski, R. F.; Ed.; *Density Functional Theory I: Functionals and Effective Potentials*; Springer Verlag: Berlin, 1996.
[9] Nalewajski, R. F.; Ed.; *Density Functional Theory II: Relativistic and Time Dependent Extensions*; Springer Verlag: Berlin, 1996.
[10] Nalewajski, R. F.; Ed.; *Density Functional Theory III: Interpretation, Atoms, Molecules and Clusters*; Springer Verlag: Berlin, 1996.
[11] Nalewajski, R. F.; Ed.; *Density Functional Theory IV: Theory of Chemical Reactivity*; Springer Verlag: Berlin, 1996.
[12] Putz, M. V. *Contributions within Density Functional Theory with Applications in Chemical Reactivity Theory and Electronegativity*; Dissertation.com: Parkland, FL, 2003 (www.dissertation.com/book.php?method=ISBN&book=1581121849).
[13] Murphy, L. R.; Meek, T. L.; Allred, A. L.; Allen, L. C. *J. Phys. Chem.* A 2000, 104, 5867-5871.

[14] Pauling, L. *J. Am. Chem. Soc.* 1932, 54, 3570-3582.
[15] Ogilvie, J. F. In *Conceptual Trends in Quantum Chemistry*; Kryachko, E. S.; Calais, J. L.; Eds.; Kluwer Academic Publishers: Dordrecht, NL, 1994, pp 171-198.
[16] Einstein, A. *Ann. Phys.* 1916, 49, 769-822.
[17] Hughes, R. I. G. *The Structure and Interpretation of Quantum Mechanics*; Harvard University Press: Cambridge, Massachusetts, 1989.
[18] Steeb, W.-H.; Hardy, Y. *Problems & Solutions in Quantum Computing & Quantum Information*; *World Scientific*: New Jersey, USA, 2004.
[19] Pauling, L. The *Nature of The Chemical Bond*; Cornell University Press: Ithaca, New York, USA, 1960.
[20] Komorowski, J. *Physique* 1983, 44, C3/1211-1214.
[21] Komorowski, L. *Chem. Phys.* 1983, 76, 31-43.
[22] Komorowski, L. *Chem. Phys. Lett.* 1983, 103, 201-204.
[23] Komorowski, L. *Z. Naturforsch.* 1987, 42a, 767-773.
[24] Komorowski, L. *Chem. Phys. Lett.* 1987, 134, 536-540.
[25] Komorowski, L. *Chem. Phys.* 1987, 114, 55-71.
[26] Komorowski, L.; Boyd, S. L.; Boyd, R. J. *J. Phys. Chem.* 1996, 100, 3448-3453.
[27] Balbás, L. C.; Alonso, J. A.; Las Heras, E. *Mol. Phys.* 1983, 48, 981-988.
[28] Bratsch, S. G. *J. Chem. Educ.* 1984, 61, 588-589.
[29] Bratsch, S. G. *J. Chem. Educ.* 1985, 62, 101-103.
[30] de Amorim, A. O. *Theoret. Chim. Acta.* 1981, 59, 551-553.
[31] Gáspár, R.; Nagy, Á. Coll. Czech. *Chem. Commun.* 1988, 53, 2017-2022.
[32] Gasteiger, J.; Marsili, M. *Tetrahedron* 1980, 36, 3219-3288.
[33] Klopman, G. *J. Am. Chem. Soc.* 1965, 87, 3300-3303.
[34] Magnusson, E. *Aust. J. Chem.* 1988, 41, 827-837.
[35] Menegon, G.; Shimizu, K.; Farah, J. P. S.; Dias, L. G.; Chaimovich, H. *Phys. Chem. Chem. Phys.* 2002, 4, 5933-5936.
[36] Mortier, W. J.; Ghosh, S. K.; Shankar, S. *J. Am. Chem. Soc.* 1986, 108, 4315-4320.
[37] Mortier, W. J.; Van Genechten, K.; Gasteiger, J. *J. Am. Chem. Soc.* 1985, 107, 829- 835.
[38] Mullay, J. *J. Am. Chem. Soc.* 1986, 108, 1770-1775.
[39] Mullay, J. *J. Comp. Chem.* 1988, 9, 399-405.
[40] Mullay, J. *J. Comp. Chem.* 1988, 9, 764-770.
[41] Mullay, *Propellants Explos.* 1987, 12, 60-63.
[42] Nalewajski, R. F. *J. Phys. Chem.* 1989, 93, 2658-2666.
[43] Parr, R. G.; Bartolotti, L. J. *J. Am. Chem. Soc.* 1982, 104, 3801-3803.

References

[44] Ray, N. K.; Samuels, L.; Parr, R. G. *J. Chem. Phys.* 1979, 70, 3680-3684.
[45] Sanderson, R. T. *Inorg. Chem.* 1986, 25, 1858-1862.
[46] Sanderson, R. T. *Inorg. Chem.* 1986, 25, 3518-3522.
[47] Sanderson, R. T. *J. Chem. Educ.* 1988, 65, 112-118.
[48] Sanderson, R. T. *J. Chem. Educ.* 1988, 65, 227-231.
[49] Van Genechten, K. A.; Mortier, W. J.; Geerlings, P. *J. Chem. Phys.* 1987, 86, 5063-5071.
[50] York, D. M.; Yang, W. *J. Chem. Phys.* 1996, 104, 159-172.
[51] Mulliken, R. S. *J. Chem. Phys.* 1934, 2, 782-793.
[52] Parr, R. G.; Donnelly, R. A.; Levy, M.; Palke, W. E. *J. Chem. Phys.* 1978, 68, 3801-3807.
[53] Berkowitz, M. *J. Am. Chem. Soc.* 1987, 109, 4823-4825.
[54] Baekelandt, B. G., Cedillo, A.; Parr, R. G. *J. Phys.Chem.* 1995, 103, 8548-8556.
[55] Ayers, P. W.; Parr, R. G. *J. Am. Chem. Soc.* 2000, 122, 2010-2018.
[56] Ayers, P. W.; Levy, M. *Theor. Chem. Acc.* 2000, 103, 353-360.
[57] Ayers, P. W.; Parr, R. G. *J. Am. Chem. Soc.* 2001, 123, 2007-20017.
[58] Ayers, P. W. *Theor. Chem. Acc.* 2001, 106, 271-279.
[59] Del Re, G. *Gazzetta Chim. It.* 1983, 113, 695-703.
[60] Del Re, G. *J. Chem. Soc. Soc. Trans.* 1981, 77, 2067-2076.
[61] Matsunaga, Y. Bull. *Chem. Soc. Japan* 1969, 42, 2170-2173.
[62] Schmidt, P. C.; Böhm, M. C. *Ber. Bunsenges. Phys. Chem.* 1983, 87, 925-932.
[63] Berkowitz, M.; Ghosh, S. K.; Parr, R. G. *J. Am. Chem. Soc.* 1985, 107, 6811-6814
[64] Pearson, R. G. *J. Am. Chem. Soc.* 1988, 110, 7684-7690.
[65] Gázquez, J. L.; Martínez, A.; Méndez, F. *J. Phys. Chem.* 1993, 97, 4059-4063.
[66] Parr, R. G.; Gázquez, J. L. *J. Phys. Chem.* 1993, 97, 3939-3940.
[67] Parr, R. G.; Zhou, Z. *Acc. Chem. Res.* 1993, 26, 256-258.
[68] Pearson, R. G. *Chemical Hardness*; Wiley-VCH: Weinheim, 1997.
[69] Senthilkumar, K.; Ramaswamy, M.; Kolandaivel, P. *Int. J. Quantum Chem.* 2001, 81, 4-10.
[70] Kolandaivel, P; Mahalingam, T.; Sugandhi, K. *Int. J. Quantum Chem.* 2002, 86, 368-375.
[71] Chandrakumar, K. R. S.; Pal, S. *J. Phys. Chem.* A 2002, 106, 11775-11781.
[72] Chandrakumar, K. R. S.; Pal, S. *J. Phys. Chem.* A 2002, 106, 5737-5744.
[73] Chandrakumar, K. R. S.; Pal, S. *J. Phys. Chem.* A 2003, 107, 5755-5762.
[74] Putz, M. V.; Russo, N.; Sicilia, E. *J. Comput. Chem.* 2004, 25, 994-1003.

[75] Putz, M. V. *Int. J. Quantum Chem.* 2006, 106, 361-389.
[76] Guo, Y.; Whitehead, M. A. *Phys. Rev.* A 1989, 39, 28-33.
[77] Nesbet, R. K. *Phys. Rev.* A 1997, 56, 2665-2668.
[78] Orsky, A. R.; Whitehead, M. A. *Can. J. Chem.* 1987, 65, 1970-1979.
[79] Ghanty, T. K.; Ghosh, S. K. *J. Am. Chem. Soc.* 1994, 116, 3943-1948.
[80] Ghanty, T. K.; Ghosh, S. K. *J. Phys. Chem.* 1996, 100, 17429-17433.
[81] Ghosh, D. C.; Biswas, R. *Int. J. Mol. Sci.* 2002, 3, 87-113.
[82] Putz, M. V.; Russo, N.; Sicilia, E. *J. Phys. Chem.* A 2003, 107, 5461-5465.
[83] Ghosh, D. C. *J. Theor. Comp. Chem.* 2005, 4, 21-23.
[84] Klopman, G. *J. Am. Chem. Soc.* 1964, 86, 1463-1469.
[85] Klopman, G. *J. Chem. Phys.* 1965, 43, S124-S129.
[86] Klopman, G. *J. Am. Chem. Soc.* 1968, 90, 223-234.
[87] Hohenberg, P.; Kohn, W.; *Phys. Rev.* 1964, 136, B684-B871.
[88] Kohn, W.; Sham, L. *J. Phys. Rev.* 1965, 140, A1133-A1138.
[89] Levy, M. *Proc. Natl. Acad. Sci. USA* 1979, 76, 6062-6065.
[90] Keller, J. *Int. J. Quantum Chem.* 1986, 20, 767-768.
[91] Davidson, E. R. *Phys. Rev.* A 1990, 42, 2539-2541.
[92] Kryachko, E. S.; Ludeña, E. V. *Phys. Rev.* A 1991, 43, 2179-2192.
[93] Kryachko, E. S.; Ludeña, E. V. *Phys. Rev.* A 1991, 43, 2194-2198.
[94] Ernzerhof, M. *Phys. Rev.* A 1994, 49, 76-79
[95] Massa, L. *Int. J. Quantum Chem.* 2002, 90, 291-293.
[96] Parr, R. G.; Pearson, R. G. *J. Am. Chem. Soc.* 1983, 105, 7512-7516.
[97] Pearson, R. G. *Inorg. Chem.* 1988, 27, 734-740.
[98] Pearson, R. G. *J. Am. Chem. Soc.* 1985, 107, 6801-6806.
[99] Pearson, R. G. *J. Org. Chem.* 1989, 54, 1423-1430.
[100] Pearson, R. G. *Proc. Natl. Acad. Sci. USA* 1986, 83, 8440-8441.
[101] Liu, G.-H.; Parr, R. G. *J. Am. Chem. Soc.* 1995, 117, 3179-3188.
[102] Böhm, M. C.; Schmidt, P. C. *Ber. Bunsenges. Phys. Chem.* 1986, 90, 913-919.
[103] De Proft, F.; Geerlings, P. *J. Chem. Phys.* 1997, 106, 3270-3279.
[104] Gázquez, J. L.; Ortiz, E. *J. Chem. Phys.* 1984, 81, 2741-2748.
[105] Robles, J.; Bartolotti, L. J. *J. Am. Chem. Soc.* 1984, 106, 3723-3727.
[106] Roy, R.; Chandra, A. K.; Pal, S. *J. Phys. Chem.* 1994, 98, 10447-10450.
[107] Berkowitz, M.; Parr, R. G. *J. Chem. Phys.* 1988, 88, 2554-2557.
[108] De Proft, F.; Liu, S.; Parr, R. G. *J. Chem. Phys.* 1997, 107, 3000-3006.
[109] Garza, J.; Robles, *J. Phys. Rev.* A 1993, 47, 2680-2685.
[110] Harbola, M. K.; Chattaraj, P. K.; Parr, R. G. *Israel J. Chem.* 1991, 31, 395-402.

[111] Nalewajski, R. F.; Korchowiec, J. *Int. J. Quantum Chem.* 1988, 22, 349-366.
[112] Putz, M.V.; Russo, N.; Sicilia, E. *Theor. Chem. Acc.* 2005, 114, 38-45.
[113] Reed, J. L. *J. Phys. Chem.* 1981, 85, 148-153.
[114] Reed, J. L. *J. Phys. Chem.* 1994, 98, 10477-10483.
[115] Zhang, Y. *Inorganic* 1982, 21, 3886-3889.
[116] Parr, R. G.; Chattaraj, P. K. *J. Am. Chem. Soc.* 1991, 113, 1854-1855.
[117] Chattaraj, P. K.; Lee, H.; Parr, R. G. *J. Am. Chem. Soc.* 1991, 113, 1855-1856.
[118] Böhm, M. C.; Sen, K. D.; Schmidt, P. C. *Chem. Phys. Lett.* 1981, 78, 357-360.
[119] Chattaraj, P. K.; Liu, G. H.; Parr, R. G. *Chem. Phys. Lett.* 1995, 237, 171-176.
[120] Cedillo, A.; Chattaraj, P. K.; Parr, R. G. *Int. J. Quantum Chem.* 2000, 77, 403-407.
[121] Nguyen, L. T.; De Proft, F.; Amat, M. C.; Van Lier, G.; Fowler, P. W.; Geerlings, P. *J. Phys. Chem.* A 2003, 107, 6837-6842.
[122] van Hooydonk, G. *J. Mol. Struct.* (Theochem) 1986, 138, 361-376.
[123] Smith, D. W. *J. Phys. Chem.* A 2002, 106, 5951-5952.
[124] Harbola, M. K. *Proc. Natl. Acad. Sci.* USA 1992, 89, 1036-1039.
[125] Ferreira, R.; de Amorim, A. O. *Theoret. Chim. Acta.* 1981, 58, 131-136.
[126] Feynman, R. P. *Phys. Rev.* 1939, 56, 340-343.
[127] Cong, Y.; Yang, Z.-Z.; Wang, C.-S.; Liu, X.-C.; Bao, X-H. *Chem. Phys. Lett.* 2002, 357, 59-64.
[128] Deb, B. M. *Rev. Mod. Phys.* 1973, 45, 22-43.
[129] Levy, M.; Perdew, J. P. *Phys. Rev. A* 1985, 32, 2010-2021.
[130] Koopmans, T. *Physica* 1934, 104-113.
[131] Gordy, W. *J. Chem. Phys.* 1946, 14, 305-320.
[132] Iczkowski, R. P.; Margrave, J. L. *J. Am. Chem. Soc.* 1961, 83, 3547-3551.
[133] Hinze, J.; Jaffé, H. H. *J. Am. Chem. Soc.* 1962, 84, 540-546.
[134] Hinze, J.; Jaffé, H. H. *Can. J. Chem.* 1963, 41, 1315-1328.
[135] Hinze, J.; Whitehead, M. A.; Jaffé, H. H. *J. Am. Chem. Soc.* 1963, 85, 148-154.
[136] Zhang, Y. *Inorganic* 1982, 21, 3889-3893.
[137] Tachibana, A. *Int. J. Quantum Chem.* 1987, 21, 181-190.
[138] Tachibana, A.; Parr, R. G. *Int. J. Quantum Chem.* 1992, 41, 527-555.
[139] Tachibana, A. *Theor. Chem. Acc.* 1999, 102, 188-195.
[140] Tachibana, A.; Nakamura, K.; Sakata, K.; Morisaki, T. *Int. J. Quantum Chem.* 1999, 74, 669-679.

[141] Pearson, R. G. *J. Am. Chem. Soc.* 1963, 85, 3533-3539.
[142] 1Pearson, R. G. *Inorg. Chem.* 1972, 11, 3146.
[143] 1Pearson, R. G. *J. Chem. Educ.* 1987, 64, 561-567.
[144] Pal, S.; Chandrakumar, K. R. S. *J. Am. Chem. Soc.* 2000, 122, 4145-4153.
[145] Chattaraj, P. K.; Maiti, B. *J. Am. Chem. Soc.* 2003, 125, 2705-2710.
[146] Parr, R. G.; Yang, W. *J. Am. Chem. Soc.* 1984, 106, 4049-4050.
[147] Pearson, R. G. *J. Am. Chem. Soc.* 1988, 110, 2092-2097.
[148] Ponti, A. *J. Phys. Chem.* A 2000, 104, 8843-8846.
[149] Drago, R. S.; Wayland, B. B. *J. Am. Chem. Soc.* 1965, 87, 3571-3577.
[150] Pérez, P.; Andrés, J.; Safont, V. S.; Tapia, O.; Contreras, R. *J. Phys. Chem.* A 2002, 106, 5353-5357.
[151] Drago, R. S.; Kabler, R. A. *Inorg. Chem.* 1972, 11, 3144-3145.
[152] Jolly, W. L.; Illige, J. D.; Mendelsohn, H. *Inorg. Chem.* 1972, 11, 869-872.
[153] McMillin, D. R.; Drago, R. S. *Inorg. Chem.* 1972, 11, 872-879.
[154] Nalewajski, R. F. *J. Am. Chem. Soc.* 1984, 106, 944-945.
[155] Méndez, F.; Gázquez, J. L. *J. Am. Chem. Soc.* 1994, 116, 9298-9301.
[156] Li, Y.; Evans, J. N. S. *J. Am. Chem. Soc.* 1995, 117, 7756-7759.
[157] Drago, R. S.; Wong, N.; Ferris, D. C. *J. Am. Chem. Soc.* 1991, 113, 1970-1977.
[158] Kohn, W.; Becke, A. D.; Parr, R. G. *J. Phys. Chem.* 1996, 100, 12974-12980.
[159] Dreizler, R. M.; Gross, E. K. U. *Density Functional Theory- An Approach to the Quantum Many-Body Problem*; Springer-Verlag: Berlin, 1990.
[160] Feil, D. *J. Mol. Struct.* (Theochem) 1992, 255, 221-239.
[161] Labanowski, J. K.; Dammkoehler, R. A.; Motoc, I. *J. Comp. Chem.* 1989, 10, 1016-1030.
[162] Lackner, K. S.; Zweig, G. *Phys. Rev.* D 1983, 28, 1671-1691.
[163] Manoli, S.; Whitehead, M. A. *J. Chem. Phys.* 1984, 81, 841-846.
[164] Matcha, R. L. *J. Am. Chem. Soc.* 1983, 105, 4859-4862.
[165] Neng-Wu, Z.; Guo-Sheng, Li *J. Phys. Chem.* 1994, 98, 3964-3966.
[166] Ohwanda, K. *Polyhedron* 1984, 3, 853-859.
[167] Putz, M. V.; Chiriac, A.; Mracec, M. *Rev. Roum. Chim.* 2001, 46, 1175-1181.
[168] Levy, M. *Phys. Rev.* A 1982, 26, 1200-1208.
[169] Politzer, P.; Lane, P.; Concha, M. C. *Int. J. Quantum Chem.* 2002, 90, 459-463.
[170] Politzer, P; Murray, J. S. *Theor. Chem. Acc.* 2002, 108, 134-142.
[171] Torrent-Sucarrat, M.; Luis, J. M.; Duran, M.; Solà, M. *J. Mol. Struct.* (Theochem) 2005, 727, 139-148.

[172] Slater, J. C. *Phys. Rev.* 1930, 36, 57-64.
[173] McWeeny, R. *Coulson's Valence,* Third Edition; Oxford University Press: Oxford, UK, 1979.
[174] Bransden, B. H.; Joachain, C. J. *Physics of Atoms and Molecules*; Longman: London, UK, 1983.
[175] Pettifor, D. G. *Bonding and Structure of Molecules and Solids*; Clarendon Press: Oxford, UK, 1995.
[176] Gillespie, R. J.; Popelier, P. L. A. *Chemical Bonding and Molecular Geometry*; Oxford University Press: New York, USA, 2001.
[177] Roth, R.; Feldmeier, H. *Phys. Rev. A* 2002, 65, 021603(R)/1-4.
[178] Viverit, L. *Phys. Rev. A* 2002, 66, 023605/1-5.
[179] Gordy, W.; Thomas, W. J. O. *J. Chem. Phys.* 1959, 24, 439-444.
[180] Bartolotti, L. J.; Gadre, S. R.; Parr, R. G. *J. Am. Chem. Soc.* 1980, 102, 2945-2948.
[181] Boyd, R. J.; Edgecombe, K. E. *J. Am. Chem. Soc.* 1988, 110, 4182-4186.
[182] Boyd, R. J.; Markus, G. E. *J. Chem. Phys.* 1981, 75, 3585-3588.
[183] De Proft, F.; Geerlings, P. *J. Phys. Chem. A* 1997, 101, 5344-5346.
[184] Gáspár, R.; Nagy, Á. *Acta Phys. Hung.* 1988, 64, 405-416.
[185] Bader, R. F. W. *Atoms in Molecules*; Clarendon Press: Oxford, UK, 1990.
[186] Daudel, R.; Leroy, G.; Peeters, D.; Sana, M. *Quantum Chemistry*; John Wiley & Sons: New York, USA, 1983.

INDEX

A

absolute electronegativity, 32, 33, 36
acid-base reaction, 71, 73
Allen, 51, 81
atomic scale, 70
atoms-in-molecules, 1, 3, 5
aufbau principle, 8, 51, 56
averaged hardness, 64

B

Berzelius, 1
bonding, 1, 2, 5, 7, 8, 9, 10, 11, 12, 14, 15, 16, 17, 19, 20, 29, 38, 39, 40, 44, 50, 52, 61, 62, 63, 71, 72, 73, 76, 77, 79, 81
bonding principles, 61
bonding scenario, 38, 39, 63, 71, 77
Bose-Einstein, 58, 59, 60
bosonic chemical action hardness, 59
bosonic density functionals, 41
bosonic hardness, 43, 44, 71
bosonic state, 40

C

charge transfer, 7, 8, 13, 39
chemical action, 20, 22, 23, 25, 26, 27, 28, 29, 31, 32, 33, 38, 40, 41, 42, 44, 46, 47, 48, 50, 53, 54, 55, 56, 57, 58, 59, 63, 65, 67, 68, 69, 70, 73, 77
chemical action principle, 27
chemical electronegativity, 33, 34, 36
chemical equilibrium, 12, 65
chemical force, 6, 11, 12, 27, 65, 76
chemical hardness, 6, 8, 32, 38, 40, 41, 42, 53, 58, 63, 65, 67, 68, 69, 73
chemical potential, 5, 6, 8, 10, 11, 12, 14, 21, 24, 25, 26, 31, 33, 37, 52, 66, 76, 77
chemical potential inequality principle, viii
chemical space, 15, 16, 71, 72
covalent, 7, 12, 14

D

delta-Dirac, 35
density functional softness theory, 34, 59
density functional theory, 8, 19, 41, 76
DFT, 18, 19, 20, 21, 22, 23, 24, 25, 31, 33, 35, 46, 50
diatomic molecule, 61, 63

E

effective nuclear charge, 46, 50

Index

electronegativity, 1, 2, 3, 4, 5, 6, 7, 8, 10, 11, 12, 13, 14, 15, 16, 18, 19, 20, 21, 22, 24, 25, 28, 29, 31, 32, 33, 34, 35, 36, 37, 38, 39, 40, 41, 42, 43, 44, 45, 48, 50, 51, 52, 53, 54, 55, 56, 58, 59, 60, 61, 62, 63, 64, 65, 66, 68, 70, 71, 72, 73, 76, 77
electronegativity equalization principle, 8, 12, 14, 25, 61
electronic affinity, 28, 32
electronic density, 20, 21, 23, 24, 26, 27, 29, 36, 50, 77
electronic fluctuations, 56
exchange and correlations, 23, 40
exchange-correlation effects, 51

F

Fermi, 19, 41
fermionic-bosonic limiting state, 73
fermionic-bosonic mixtures, vii
finite-difference, 59

G

gauge reactions, 7
global softness, 17, 33, 34, 35

H

hard and soft ordering, 70
hard base, 15, 18, 66
hardness equalization, 13, 16, 17, 61
hard-soft, 14, 17, 69, 71
Hartree-Fock, 67, 68, 69, 70, 71, 72
Heidegger, 76
Hohenberg, 19, 20, 23, 46, 84
HOMO, 9, 14, 28, 52, 53, 56, 57, 58, 59
HSAB, 12, 13, 14, 16, 17, 18, 19, 29, 36, 39, 58, 61, 65, 70

I

ionization potential, 3, 6, 28, 32, 45

K

Kant, 75, 77
Kohn, 19, 20, 23, 46, 84, 86

L

Lewis, 7, 14, 65, 66, 70, 72, 77, 81
limited chemical action hardness, 38
local chemical work, 27
local softness, 33, 35, 36, 37, 59
local-nonlocal picture, 34
LUMO, 9, 14, 28, 52, 53, 56, 57, 58, 59

M

maximum hardness principle, viii
Maxwell, 31
Mel Levy, 23
Mendeleyev, 51
MH principle, 16, 17
molecular orbital theory, 8, 9
Moseley, 51
Mulliken, 3, 4, 5, 19, 58, 83

N

Newton, 75
Noica, 75
nonlocal effects, 40, 42
N-representability, 24

O

occupation number, 7
orbital exponent, 46, 49, 50
orbitals, 1, 9, 10, 14, 23, 27, 31, 39

Index

P

parabolic form, 14
Parr, 19, 81, 82, 83, 84, 85, 86, 87
partial charges, 5
Pauling, 2, 3, 5, 19, 58, 63, 76, 82
Pearson, 19, 65, 66, 72, 83, 84, 86
Pearson classification, 65
People, 19
Periodic System, 8
periodicity of elements, 59
Plato, 8
polarizability, 15, 17
polyatomic hardness, 65

Q

quantum chemistry, 2, 19, 20, 77
quantum fluctuations, 12, 13, 58, 77

R

reaction energies, 14, 71
reactivity, 1, 4, 5, 6, 7, 8, 10, 11, 12, 13, 14, 16, 17, 18, 19, 20, 21, 22, 23, 29, 35, 37, 39, 40, 44, 45, 52, 56, 58, 60, 65, 66, 71, 72, 76, 77
response function, 35, 38, 47

S

Sham, 19, 84
Slater, 19, 46, 49, 50, 87

soft acid, 15, 18, 66
soft base, 15, 18
soft-hard, 14, 17, 18, 71, 72
softness kernel, 34, 35, 36
soft-soft, 14, 17, 18, 71
Sommerfeld, 19
superconductivity, 58

T

Thomas, 19, 41, 87
total energy, 5, 6, 10, 20, 21, 22, 24, 25, 26, 28, 29, 34, 45, 62

U

universal invariant, 63

V

$V(x)$-representability, 24
valence, 1, 8, 10, 19, 21, 22, 23, 29, 32, 34, 45, 46, 47, 49, 50, 51, 56, 61
valence state, 10, 21
variational principle, 11, 21, 22, 24, 25, 40, 56
virial relation, 58
virial theorem, 26, 27

W

wave-function, 2, 20